gy

TERTIARY LEVEL BIOLOGY

A series covering selected areas of biology at advanced undergraduate level. While designed specifically for course options at this level within Universities and Polytechnics, the series will be of great value to specialists and research workers in other fields who require a knowledge of the essentials of a subject.

Recent titles in the series:

Mammal Ecology	Delany
Virology of Flowering Plants	Stevens
Evolutionary Principles	Calow
Saltmarsh Ecology	Long and Mason
Tropical Rain Forest Ecology	Mabberley
Avian Ecology	Perrins and Birkhead
The Lichen-Forming Fungi	Hawksworth and Hill
Social Behaviour in Mammals	Poole
Physiological Strategies in Avian Biology	Philips, Butler and Sharp
An Introduction to Coastal Ecology	Boaden and Seed
Microbial Energetics	Dawes
Molecule, Nerve and Embryo	Ribchester
Nitrogen Fixation in Plants	Dixon and Wheeler
Genetics of Microbes (2nd edn.)	Bainbridge
Seabird Ecology	Furness and Monaghan
The Biochemistry of Energy Utilization in Plants	Dennis
The Behavioural Ecology of Ants	Sudd and Franks
Anaerobic Bacteria	Holland, Knapp and Shoesmith
An Introduction to Marine Science (2nd edn.)	Meadows and Campbell
Seed Dormancy and Germination	Bradbeer
Plant Growth Regulators	Roberts and Hooley
Plant Molecular Biology (2nd edn.)	Grierson and Covey

TERTIARY LEVEL BIOLOGY

Polar Ecology

B. STONEHOUSE, BSc, MA DPhil
Scott Polar Research Institute
Cambridge

Blackie
Glasgow and London

Published in the USA by
Chapman and Hall
New York

Blackie and Son Limited,
Bishopbriggs, Glasgow G64 2NZ
7 Leicester Place, London WC2H 7BP

Published in the USA by
Chapman and Hall
a division of Routledge, Chapman and Hall, Inc.
29 West 35th Street, New York, NY 10001–2291

First published 1989

British Library Cataloguing in Publication Data

Stonehouse, Bernard, *1926–*
Polar ecology.
1. Polar regions. Ecology
I. Title II. Series
574.5′0911

ISBN 0-216-92480-4
ISBN 0-216-92481-2 Pbk

For the USA, International Standard Book Numbers are
0-412-01701-6
0-412-01711-3 (pbk.)

Phototypesetting by Thomson Press (I) Ltd.
Printed in Great Britain by Bell & Bain (Glasgow) Ltd.

Preface

Polar Ecology is one ecologist's attempt to sum up plant, animal and environmental relationships in the polar regions. Ecology grabs ecologists in different ways. I was grabbed in Antarctica by Adelie penguins, incubating contentedly at −20°C with unmelting snow on their backs, and by minute black insects basking at 10°C in tufts of moss, while winds at −15°C swept past unheeded. Some time later I saw snow buntings sheltering under Canadian eaves at −30°C, and wondered (as I still wonder) how so tiny an organism maintains body temperature against so sharp a gradient.

The subtitle of this book, if it had one, would be '... an environmental approach'. My interests centre squarely on plants and animals, but it is their responses to the environment—the physical conditions in which they find themselves—and effects of environmental constraints on their communities, that intrigue me most. In a small book, this has left little room for other important aspects of ecology—for example, the production and process ecology that currently preoccupy field researchers, and the biogeography and evolution of polar ecosystems that still provoke argument and speculation. My approach may provide background for other aspects of ecology, both polar and world-wide. I hope that the in-text citations and end-of-chapter bibliographies will help students to find their way into the broader fields beyond.

BS

Contents

CHAPTER ONE
INTRODUCTION: POLAR ENVIRONMENTS

1.1 Polar and subpolar regions

The Arctic and Antarctic, cold regions of land and ocean at the ends of the earth, surround the North and South Poles (Figure 1.1). The Arctic is centred on an ocean basin, and includes a fringe of continental land and islands. The Antarctic includes a high, ice-covered continent, off-lying islands and a surrounding ocean. Narrow subpolar zones, the Subarctic and Subantarctic, separate polar from temperate regions in the two hemisphere. For geographical and general accounts of polar regions see Sugden (1982).

There is no general agreement on the limits of the polar and subpolar regions. The terms Arctic, Antarctic, Subarctic and Subantarctic are used in common speech, and geographers, climatologists, lawyers, administrators, politicians and ecologists all use them in slightly different ways, depending on where they draw their boundaries. Here we deal mainly with ecological regions and boundaries, but consider some of the alternatives that appear in polar literature.

Atlases generally represent polar regions badly, though most large atlases have at least one small-scale map of each. The only English-language atlas covering both polar regions well is one produced by the US Central Intelligence Agency (CIA, 1978).

1.1.1 *Boundaries*

Administrators and geographers favour straight-line boundaries that can be simply defined and seen on maps, for example meridians and parallels of latitude; for them the Arctic and Antarctic Circles are good polar boundaries. Climatologists use isotherms (lines joining points of equal mean temperature) to define climatic regions, but often include characteristic vegetation as well (see below). Ecologists prefer boundaries that are visible or measurable in the field, and separate plant and animal communities. Terrestrial ecologists often use boundaries between prominent plant

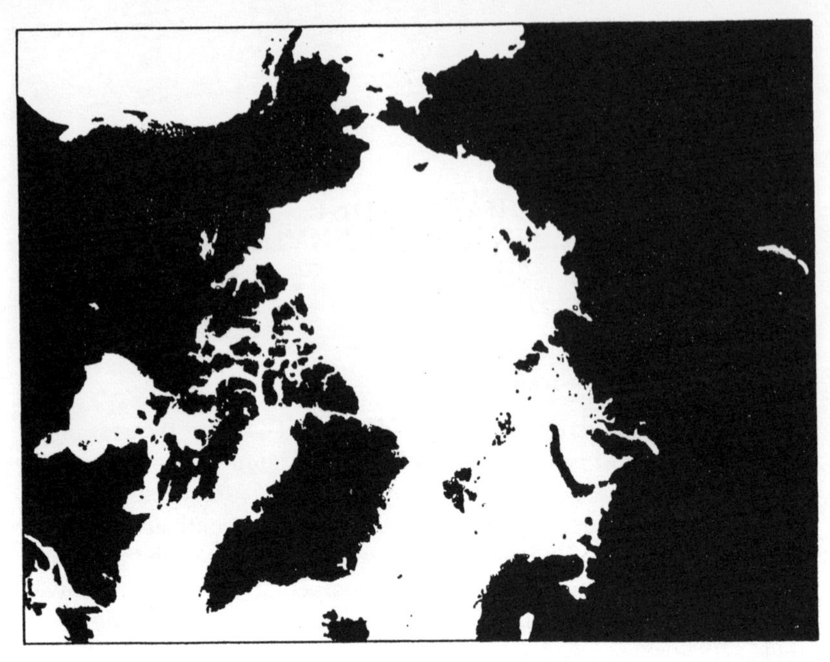

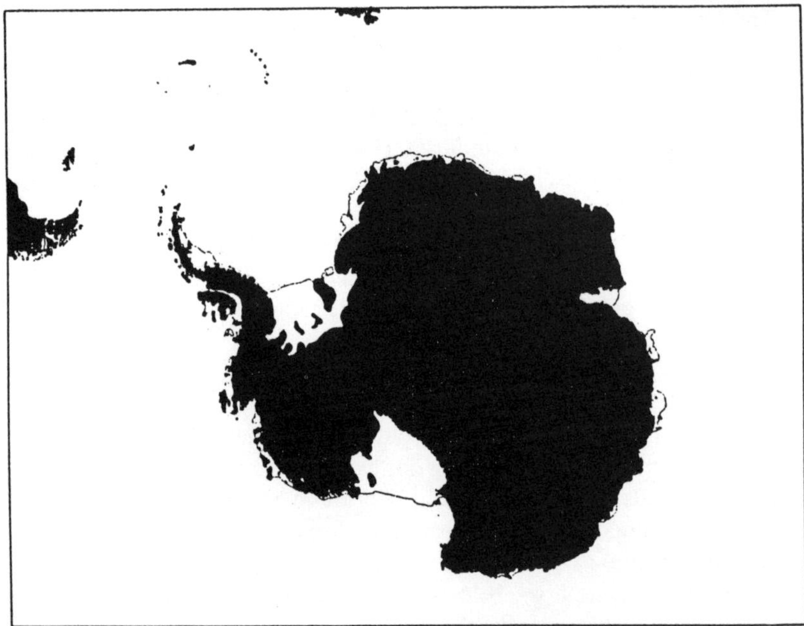

Figure 1.1 Distribution of land (black) and ocean in north (upper) and south polar regions.

communities; marine ecologists use boundaries between water masses, which separate marine plant and animal communities too. For discussion of principles underlying physiographic zoning see Govorukha and Kruchinin (1981).

Parallels of latitude. The polar circles are parallels of latitude approximately 66°32′ from the equator and 23°28′ from the poles, defined by the angle between the axis of rotation of the earth and the ecliptic, or plane of earth's orbit about the sun (Figure 1.2). The axis of rotation varies by about 1° per century in relation to the ecliptic, but is constant enough for the polar circles to be fixed on 66°32′ north and south.

Early geographers used the polar circles and the Tropics of Capricorn

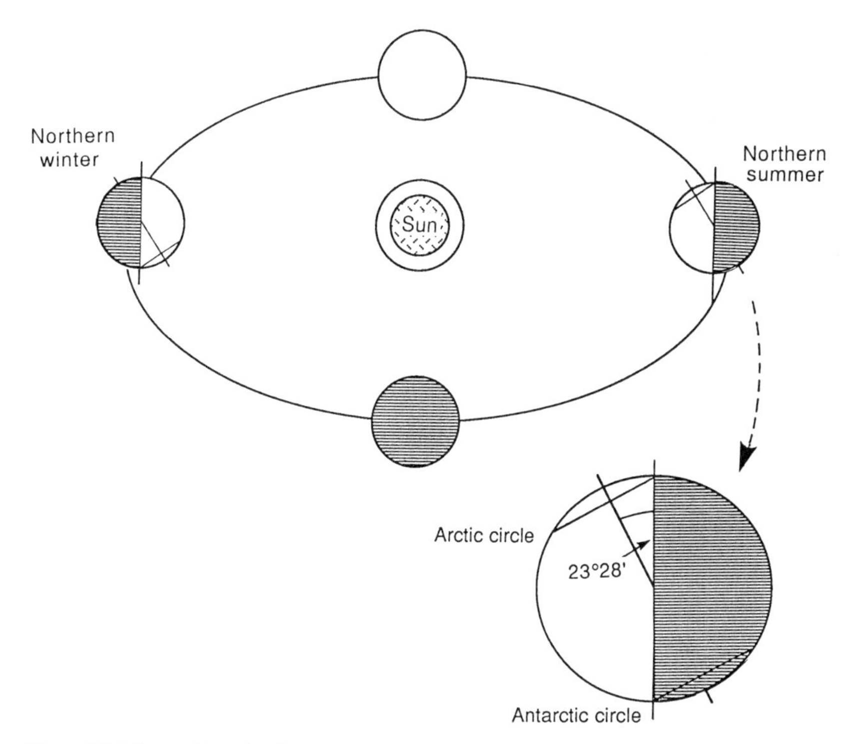

Figure 1.2 The earth's axis of rotation is tilted about 23°28′ from vertical in relation to the ecliptic or plane of rotation of the earth around the sun. Resulting from the tilt, the earth receives differential illumination during its yearly circulation around the ecliptic. Regions poleward of the Arctic and Antarctic circles are fully exposed to solar radiation across midsummer and fully shaded across midwinter.

and Cancer (parallels of latitude approximately 23°28′ on either side of the equator) to divide the world's surface into broad polar, temperate and tropical zones. To the northern cold zone that lay under Arctos, the pole star, they gave the name Arctic, and the cold zone at the opposite end of the world became the Antarctic. These usages have persisted in modern speech, and geographers and many others still find the polar circles useful in defining north and south polar regions.

Polar circles are clear on maps but not on the earth's surface. Day length and the amount of solar radiation received are ecologically important factors that vary with latitude, for solar radiation is the main source of heat for the earth's surface, the main driving force behind winds and oceanic currents, and the ultimate source of energy for all living creatures. Areas within the circles are colder than the rest of the world because they receive less solar radiation per year (Chapter 2), but there is no sharp transition at the circles themselves. On the poleward side the sun remains above the horizon on midsummer day and below it on midwinter day. The sunless period in winter and the period of continuous daylight in summer both increase from one day per year at the polar circles to six months per year at the poles. However, because there is no notable change in radiation levels at 66°32′ north or south, the circles have no ecological significance; neither plants nor animals respond to them.

Latitudinal variations in solar radiation broadly determine the global distribution of climates, and therefore of plants and animals. However, relief and other geographical factors exert strong local influences that upset simple latitudinal patterns, influence local climates, and determine where plants and animals actually live. Because of geographic differences between the northern and southern polar regions, notably the dominance of an ocean basin in the north and of a high continent in the south, the areas within the polar circles differ widely in climates and ecology. Herein lies the circles' most useful property for ecologists. Comparing climatic figures for stations close to the circles (Tables 2.7, 2.8), or indeed for any other equivalent latitudes, it becomes clear that, latitude for latitude, the southern hemisphere is colder. Within the Antarctic Circle lies little more than a high desert continent, populated only by scientists and support staff. Within the Arctic Circle are forests, cultivated ground, towns and cities, and a growing human population, currently numbering about 2 million.

That polar influences extend over a much wider area south than north was first noted by the early explorers (Mill, 1905). The French navigator Yves-Joseph de Kerguélen-Trémarec, who in 1772 glimpsed land in 49°S, a latitude similar to that of northern France, thought he had found a fertile

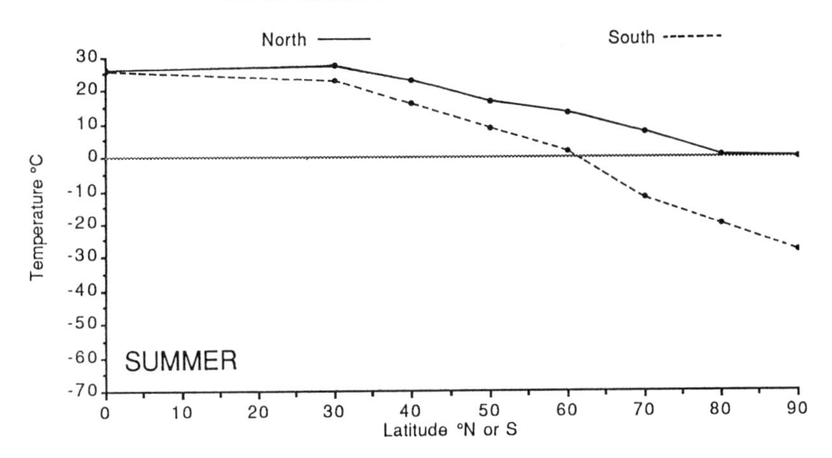

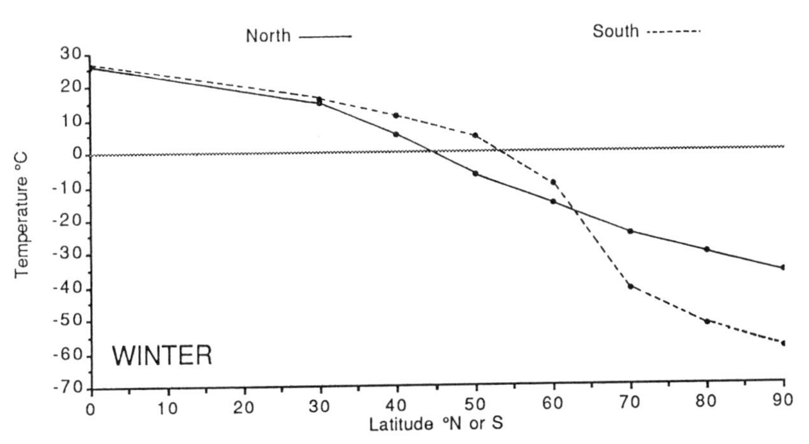

Figure 1.3 Mean air surface temperatures for midsummer and midwinter in latitudinal zones. Latitude for latitude the southern hemisphere is more stable than the northern throughout the year, reflecting its higher proportion of ocean. In summer it is generally cooler; in winter it is warmer in mid-latitudes but much colder toward the pole. Data from Markov and others (1970), augmented.

south continent. Returning a year later with an armada of colonists, he discovered the reality—the archipelago of cold, stormbound, glaciated islands now called Iles Kerguelen. Captain James Cook in 1775 recorded that South Georgia, in a southern latitude equivalent to that of his native northern Britain, had a much colder climate, snow-covered at the height of summer with glaciers flowing into the sea; '... not a tree was to be seen,' he wrote, 'nor a shrub even big enough to make a toothpick' (Cook, 1784). Other parallels of latitude find favour for similar reasons—because they are readily referred to, easily understood, and point similarities or differences between the two hemispheres. Marine insurers increase premiums when ships cross the 60th parallel north or south. Lawyers have defined the boundary of the Antarctic Treaty area (Chapter 7) in 60°S; this serves many practical purposes, though some ecologically-based provisions of the Treaty transcend the boundary to cover islands and stretches of oceans beyond it.

Climatic boundaries. For climatologists, polar regions are areas where polar climates prevail. Climate classifications are based primarily on temperature and precipitation. Some incorporate other climatic elements, for example evaporation, that have a direct bearing on growth of vegetation. Others bring in types of vegetation, regarding them as diagnostic of climate; these are the ones that ecologists find most useful; for a discussion of alternative systems and their polar relevance see Tuhkanen (1980).

W. Köppen's system of classification is the one most generally accepted (Köppen and Geiger, 1936); Thornthwaite (1948) elaborated it, taking into account additional factors of ecological significance, but the simplicity of the original still has wide appeal. Köppen distinguished five major types of world climate, defined primarily by temperature and vegetation:

A: equatorial and tropical rain climates
B: dry climates of the arid zone
C: temperate climates of the broad-leafed forest zone
D: cold temperate forest climates
E: treeless polar climates

Each of the warmer climates is divided into subtypes according to amount and time of incidence of rainfall. The main criterion for polar climates is that the mean temperature for every month is below 10°C (50°F); polar climatic regions so defined are enclosed by the 10° isotherm for the warmest summer month, normally July in the northern hemisphere and January in the south (Figures 1.4, 1.5).

For the Arctic this isotherm encircles the Arctic Ocean and includes Greenland, Svalbard, the northern two-thirds of Iceland, and most of the northern coasts and offlying islands of the Soviet Union, Canada and Alaska. In the western Atlantic Ocean it is pushed anomalously northward by air masses associated with the North Atlantic Drift or Gulf Stream, to

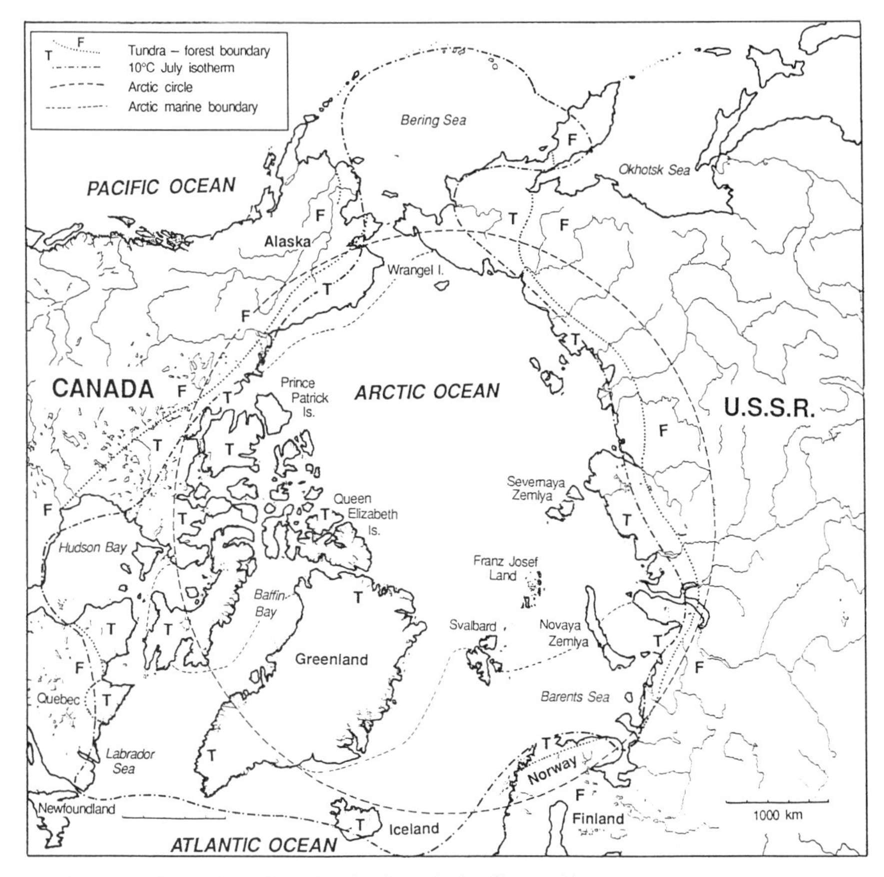

Figure 1.4 The arctic region, showing boundaries discussed in text.

exclude all but the northernmost extremities of Scandinavia. In the eastern Atlantic sector the isotherm loops southward, influenced by cold sea and air currents from the polar basin, to bring northeastern Labrador, northern Quebec and Hudson Bay into the Arctic. In the northern Pacific similar currents push the isotherm southward to take in central Kamchatka and much of the Bering Sea (Figure 1.4). For the Antarctic the 10° summer isotherm encircles the whole of continental Antarctica and includes the southwestern tip of South America, Tierra del Fuego, and a wide swathe of the Southern Ocean (Figure 1.5).

These climatically-defined areas fit well with most people's concept of polar regions; they are cool in summer and bitterly cold in winter, with

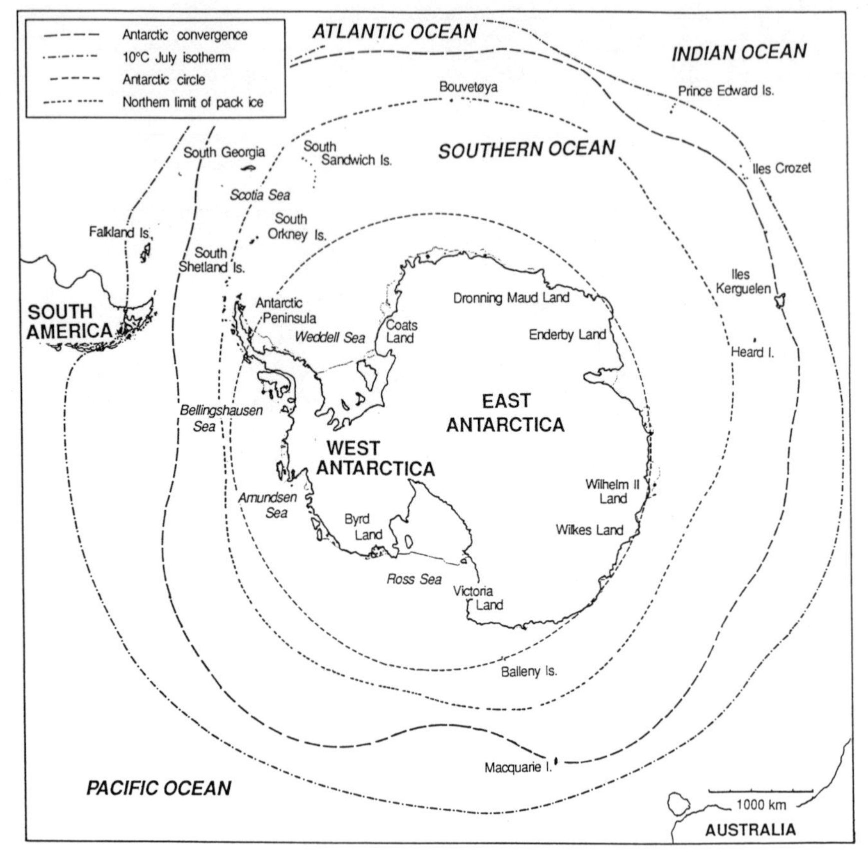

Figure 1.5 The antarctic region, showing boundaries discussed in text. Note that, because polar cold spreads further in the south than in the north, this map has to be on a slightly smaller scale than Fig. 1.4.

snow more plentiful than rain, and water and ground frozen for much of the year.

Some ecologists prefer the 10°C summer isotherm to any other polar boundary because it is equally valid over land and ocean, and on land corresponds fairly closely to the treeline (see below). Further, it is the same for both hemispheres and provides a basis for comparison between them, enclosing a much wider area in the south than in the north. Polar ecologists who are more directly concerned with vegetation than with temperature have reservations about it; especially in the Arctic, the 'treeless' aspect of Köppen's definition is more significant for them than the related isotherm.

Climatology provides subpolar boundaries too; in Köppen's classification a subpolar climate is one in which the mean temperature of the coldest month is below freezing point, and the means for the four warmest months are below 10°C. Summers are long enough and warm enough to promote tree growth, but only of hardy species that can withstand the long, often bitterly cold winters that follow. Winters are in fact colder over much of the Subarctic—for example in the heartlands of northern Canada and Siberia—than in the Arctic itself (Figure 2.7). The equivalent Subantarctic zone covers a wide expanse of ocean and a scattering of small islands and archipelagos, where westerly winds blow strong and maritime influences stabilize temperature throughout the year.

The treeline. The treeline is the northern limit beyond which trees do not grow—where forests disappear and are replaced by tundra. In fact each species of tree has its own treeline, reflecting different requirements for optimum growth. What constitutes a tree is arguable (p. 87); however, in real life the limit of man-sized trees (1.5–2.0 m) may be quite sharply defined. It is sometimes clearest on hillsides, where altitude becomes an important limiting factor, and least defined in rolling country where patches of marsh and permafrost (permanently frozen subsoil) provide a soil mosaic. Most often the limit is a narrow zone called forest-tundra, where continuous forest gives way to open ground, with occasional stands of wind-sculptured and stunted trees. The transition from mainly forest to mainly tundra is usually clear in the field, in aerial photographs and in images from polar-orbiting satellites.

The treeline boundary appeals to ecologists because treeless tundra is an easily recognized biome, involving particular kinds of soil, vegetation and fauna. Arctic tundra is entirely characteristic of the Arctic and found nowhere else: alpine and antarctic tundra ecosystems are similar but not identical.

For much of its length the treeline keeps close to the 10°C summer isotherm, reflecting the strong influence of summer temperatures on tree growth. A slightly closer fit is achieved with the 'Nordenskjöld line', an isopleth connecting all places where the mean temperature of the warmest month equals $(9 - 0.1\,k)$, k being the mean temperature of the coldest month in °C (Huschke, 1959).

However, departures arise because winds, soils, topography, availability of water and other local factors also affect tree growth. Isolated stands of trees may grow well north of the general line, for example in sheltered valleys, and growth may be discouraged in unfavourable areas to the south, for example on windy maritime islands. Further, the presence or absence of trees can be plotted accurately over large areas, but isotherms are drawn from records of stations that are few and far apart. Sparsity of data defeats efforts to match the treeline closely with isotherms, or with combinations of temperature, evaporation and other variables. For the terrestrial Arctic the treeline is a useful practical boundary, and the 10°C summer isotherm a useful conceptual one, and ecologists need to be aware of both. There is no striking southern biological boundary for the Subarctic.

In the southern hemisphere the treeline again follows fairly closely the 10°C summer isotherm, cutting the southern tip of South America and Tierra del Fuego, weaving south of the oceanic islands that have trees, and north of those that do not. Few ecologists take it seriously; this is a clearly a maritime region, for which maritime boundaries are preferable.

Maritime boundaries. The water masses that make up the oceans differ from each other in temperature, salinity, electrical conductivity and density according to their origins and history (Chapter 5). Oceanographic ships now monitor constantly the water through which they are passing; a sudden change in one or more of these variables indicates that the ship has crossed a boundary between one water mass and another. Such boundaries occur in every ocean, both at the surface and in depth; each is a zone of mixing between two adjacent masses. Though they shift radically in response to long-term climatic variations, and their mean position may vary over several kilometres even within the human lifespan, from year to year the boundaries remain surprisingly stable. They are ecologically significant because water masses tend to gather their own assemblies of species, which are characteristic and do not form in other waters of different temperatures and salinities. Thus the boundaries between water masses separate distinct communities of plants and animals.

The arctic maritime boundary forms between cold, dilute surface waters from the Arctic Ocean, and warmer, saltier waters from the south. Arctic oceanic water, which is ecologically poor, spreads southward to about 63°N through the Canadian archipelago and along east Baffin Island, and to 65°N off east Greenland. It meets and over-rides mixed arctic and Atlantic water, ecologically more productive, which spreads in a wide belt from Labrador in the west to Novaya Zemlya in the east (Figure 1.4).

Off northwestern Scandinavia the belt is pushed far to the north by the North Atlantic Drift (Gulf Stream); near Svalbard the arctic maritime boundary lies north of 80°N. Similarly in the northern Pacific area warm oceanic water from the south mixes with arctic water around the Bering

Strait and is thrust northward. It spreads east as far as Amundsen Gulf along the north Alaskan coast, meeting arctic oceanic water along a boundary that lies in about 72°N; to the west it spreads almost to Wrangel Island. The subarctic maritime boundary forms between the mixed water and the warmer waters of the north Atlantic and Pacific Oceans, over a wide range of latitudes from about 44°N off Newfoundland to 68° off western Scandinavia.

For the southern polar region the most generally accepted boundary is the Antarctic Convergence (Figure 1.5), a belt of water in places only a few kilometres wide, separating antarctic surface water from slightly warmer and more saline subantarctic surface water. The convergence, detectable at the surface as a sudden sharp gradient in sea temperature (Figure 1.6), marks the line where the colder southern water, moving east and slightly northward before the prevailing winds, sinks below the northern water which is driving east and slightly southward. Currently it snakes around the world mainly between latitudes 47°S and 62°S, generally further south in the Pacific than in the Atlantic sector, and shifting slightly southward in summer and north in winter. The most widely accepted subantarctic boundary is the Subtropical Convergence, a less clearly defined belt of water that circles the earth in lower latitudes, mostly between 36° and 40°, marking the zone where subantarctic water sinks below warmer temperate water masses.

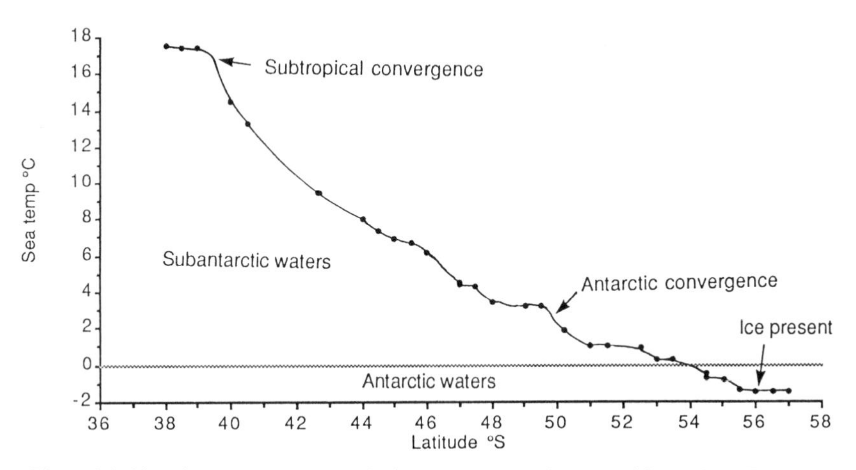

Figure 1.6 Changing sea temperature during a voyage southwestward from Cape Town to the Weddell Sea, Antarctica, early spring 1986, along a straight course. The positions of the subtropical convergence and Antarctic convergence are indicated by rapid temperature shifts.

1.1.2 *The anomaly of polar cold*

Cold polar regions are a seldom-occurring anomaly in world history. In an earth free of ice-caps, solar radiation and the circulation of atmosphere and ocean ensure that temperate climates extend into high latitudes; this seems to have been the norm over much of the earth's history. We are currently in an ice age which began with the glaciation of Antarctic during the late Tertiary, some 20–30 million years ago (Andrews, 1979; John, 1979a). The ice age before it occurred in Permo-Carboniferous times some 250 million years ago (John, 1979b), and the long period between was one of temperate rather than frigid poles. For an account of changing climates see Lamb (1972, 1977).

For polar ice-caps to form, two conditions are required—firstly a global cooling to ensure that snow becomes the prevalent form of polar precipitation, and secondly enough land close to the poles for snow to accumulate. Many possible causes have been suggested for the current ice age (Flint, 1971). A most likely combination of events was the massive uplifting of portions of the earth's surface—the orogenies that raised the Himalayas, European Alps, and the western mountains of North and South America—during the late Tertiary, coupled with the coincidence at the same time of a wandering South Pole and a wandering Antarctic continent, the latter a recent escape from the disintegration of the Gondwana supercontinent.

The orogeny raised large tracts of land above the snowline all over the world, allowing snowfall to accumulate, reflect back a proportion of incident solar radiation and start a process of world cooling. The presence of a mountainous continent in the far south, undergoing massive orogenies of its own, allowed snow to accumulate there and form a vast additional reflector. At the same time atmospheric and ocean currents in the far south changed from meridional to latitudinal, isolating Antarctica and intensifying its cold (Figure 1.7). The antarctic ice sheet had probably spread over most of the continent 10 million years ago, and reached its present proportions 4 million years ago (Figure 1.8). The Greenland ice sheet began to form some 3–4 million years ago, and the lesser ice sheets that grew in Europe, Siberia and North America followed soon after (Figure 3.1).

The ice age was punctuated by glacial periods or glacials in which intense cooling caused the ice-caps to spread, and warmer interglacials when the ice retreated. The causes of cooling and warming were not the same as those that initiated the ice age; changes in levels of solar radiation are a likely possibility. The effects have been well studied, particularly from the rock debris and fossil record left by retreating ice-sheets. During glacial periods

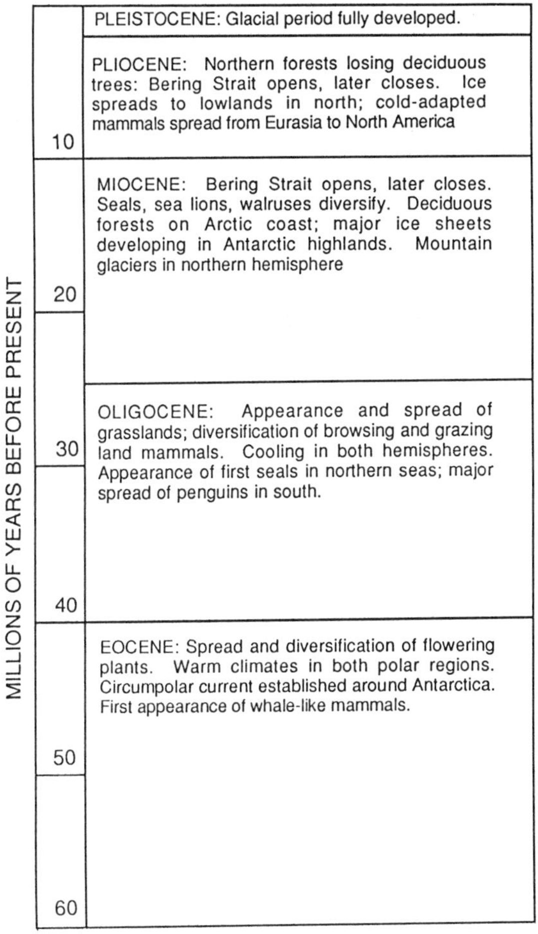

MILLIONS OF YEARS BEFORE PRESENT	
	PLEISTOCENE: Glacial period fully developed.
10	PLIOCENE: Northern forests losing deciduous trees: Bering Strait opens, later closes. Ice spreads to lowlands in north; cold-adapted mammals spread from Eurasia to North America
20	MIOCENE: Bering Strait opens, later closes. Seals, sea lions, walruses diversify. Deciduous forests on Arctic coast; major ice sheets developing in Antarctic highlands. Mountain glaciers in northern hemisphere
30 40	OLIGOCENE: Appearance and spread of grasslands; diversification of browsing and grazing land mammals. Cooling in both hemispheres. Appearance of first seals in northern seas; major spread of penguins in south.
50 60	EOCENE: Spread and diversification of flowering plants. Warm climates in both polar regions. Circumpolar current established around Antarctica. First appearance of whale-like mammals.

Figure 1.7 Events following world cooling during the past 60 million years. After Stonehouse (1972) and Imbrie and Imbrie (1979).

ice-sheets build up on land, and sea level falls as water is withdrawn. During interglacials parts of the ice-sheets disappear and sea level rises. Four major glacial periods have been identified; we are currently in a fourth interglacial which could well be followed by further cooling (John, 1979a).

This history is strongly relevant to polar ecology. Within the last few million years, long-established temperate and subtropical communities have been swept from the lands which are now polar, including

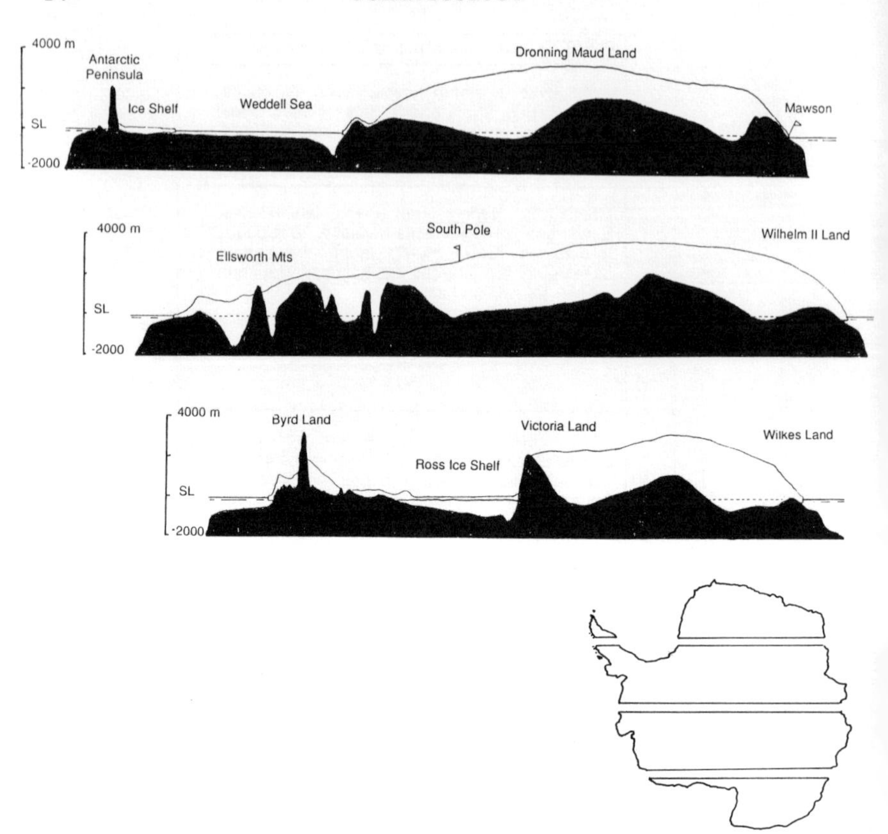

Figure 1.8 The Antarctic ice cap.

sabre-toothed tigers, palm trees and mastodons from the Arctic and beech forests, reptiles, and mammals from Antarctica (Kurtén, 1972, Nilsson, 1983). 'Beringia', a land that appeared intermittently across the Bering Strait during glacial periods of sea-level depression, allowed temperate species to travel between Eurasia and America. To this we owe much of our present terrestrial circumpolar flora and fauna (Chapter 3). One of the last species to cross Beringia was man, who invaded North America from Asia less than 100 000 years ago.

Since the ice age itself has lasted only a few million years, polar species have had little time to adapt to polar conditions. Though there is evidence at both ends of the earth for plant and animal refugia—favoured areas where a few pre-pleistocene species survived the glacial episodes—most polar species are currently still invading and adapting from neighbouring temperate regions; the paucity of polar species, in comparison with temperate or tropical species, is due partly to lack of time. Since the last glacial ended and its ice-sheets retreated only a few thousand to a few hundred years ago, most polar habitats and communities are indeed new, a point polar ecologists must always keep in mind.

1.1.3 *Maps, conventions and terms*

Polar regions are often poorly represented on world maps, usually with distorted shapes, directions and areas. Globes and polar projections give far better impressions of polar geography, and most world atlases include at least one map of each. By convention north generally appears at the top of a map and south at the bottom, but polar projections show the pole in the centre and the prime meridian running from top to bottom. This is the most useful way of visualizing both polar regions mentally.

Arctic, Subarctic, Antarctic and Subantarctic start with capital letters when they are used as the names of regions, and as part of a regional name, e.g. Arctic Ocean, Subantarctic zone. They are not usually capitalized when used adjectivally (e.g. arctic climate, antarctic fish), though editors vary in applying this rule.

Note that there is an Arctic Ocean, but the ocean surrounding Antarctica is the Southern Ocean, not the Antarctic Ocean. Both the Arctic Ocean and the Southern Ocean have fringing seas, some of which are still acquiring names. Most prominent geographical features (mountains, headlands, bays, etc.) in the north and in the Subantarctic have now been named, but new names are still being added to antarctic maps, and coastlines and other features change as new information is incorporated.

Geographical names are now generally given in the language of the country that claims the territory. Bear Island in the Arctic is more often rendered in Norwegian as Bjørnøya; in the Subantarctic French-owned Kerguelen Islands are now Archipelago Kerguelen, and Russian Franz-Joseph Land has become Zemlya Frantsa-Iosifa. Thus polar literature can be confusing; an area of interest may have no official name or several different ones. In this book the most familiar forms of names are used, with new or less familiar variants given in brackets. While ownership is generally undisputed in the north and in the Subantarctic, it is less clear in the Antarctic, where some nations dispute and others ignore territorial claims. All claims are currently shelved under the Antarctic Treaty though still jealously guarded by claimant nations.

1.2 Arctic and Subarctic

1.2.1 *Arctic topography*

Taking the treeline as the southern boundary, the Arctic region includes all of the Arctic Ocean, all of Greenland and Iceland, Svalbard, Franz Josef Land, the Soviet islands of Novaya Zemlya, Severnaya Zemlya, Novosibirsk and Wrangel, and the many islands of the Canadian arctic archipelago. On mainland Eurasia the boundary crosses Kola Peninsula—only the northern coast is treeless—and runs roughly parallel to the Siberian coast, which is virtually treeless for up to 500 km island. Most of the mountainous northeastern corner of the USSR beyond Kolyma River is treeless, as far south as the neck of Kamchatka Peninsula. In the Bering Sea St Lawrence Island, Nunivak, the Pribilov Islands, the eastern Aleutian Islands and much of the western and northern coasts of Alaska are arctic; so is a wide swathe of the north Canadian coast including northern shores of Hudson Bay, Ungava Peninsula and northern Labrador.

The 10° summer isotherm lies 100–200 km north of the treeline over much of mainland Eurasia and North America. The Arctic region as defined by this isotherm is slightly smaller, excluding the southernmost mountains of eastern Asia beyond the Kolyma, excluding western Alaska, the western Aleutians and southern Iceland, but including the northern half of Kamchatka Peninsula. By either definition the region is markedly asymmetric about the North Pole; in Eurasia it remains almost entirely within the Arctic Circle, but in eastern Canada extends south even beyond the 55th parallel.

The land. The land area is topographically varied. Greenland, an archipelago of mountainous islands, is almost completely hidden under ice. Its ice-cap, rising to 3230 m and maintained by heavy snowfall, is the largest northern remnant of the last glacial period. There are smaller caps on neighbouring Canadian islands—Baffin, Devon, Ellesmere and Axel Heiberg, among mountains over 2000 m high. Iceland by contrast is mostly ice-free, with only small glaciers; it straddles the mid-Atlantic Ridge and is volcanically lively.

Ice covers much of the rugged Svalbard archipelago and caps other islands north of Eurasia, including Novaya Zemlaya, a northern extension of the Ural Mountains. Much of the Siberian coast is low-lying and snow-free in summer; the tundra-covered shores are punctuated by deltas of the great rivers which carry silt and warm fresh water into the Arctic Ocean. High ice-covered mountain ranges dominate the peninsulas of Gory

Byrranga and Chukotskoye Nagor'ye, with many peaks above 2000 m. Alaska's rugged mountain ranges carry permanent ice, but the coastal plains are seasonally snow-free; the arcuate volcanic chain of the Aleutians, and the islands of the central Bering Sea, have only winter snow-cover. The northwest Canadian Arctic is mostly rolling upland country with thin, intermittent snow-cover for much of the year.

The ocean. The Arctic Ocean, with an area of about 10 million km^2, occupies a deep, complex basin (Chapter 4), linked to the north Pacific Ocean through Bering Strait and to the Atlantic through the Canadian archipelago and between Greenland and Scandinavia. Its centre has a permanent cover of slowly circulating, interlocking ice floes, most of them several years old and many 3–4 m thick. Surrounding the core is a zone of seasonal pack ice, and surrounding that a zone of fast ice, so-called because it is fast to the land. Fast ice and seasonal pack ice grow to about 1 m thick and disperse annually; some floes are recruited into the central pack, but most are dispersed by melting and by centrifugal currents that carry them south into the north Atlantic. Surface waters are of low salinity, especially over the Eurasian shelf where rivers contribute huge masses of warm fresh water.

Sea ice is important ecologically, providing a substrate for algae to grow and for seals to breed on. However, sea surface temperatures in its presence seldom rise above −1°C, sunlight cannot penetrate the water, and gaseous exchanges between surface and atmosphere are restricted; productivity is accordingly low. Sea ice also has a depressing effect on islands that it invests, keeping them from warming in summer and restricting their productivity.

1.2.2 *Glacial and ecological history*

Some 5–10 million years ago the climate of the Arctic changed from warm to cool-temperate; between 3 and 5 million years ago it hardened, ice investing the uplands and spreading to the plains (Figure 1.7). As cooling intensified and the ice accumulated on land, the Beringian land bridge (see above) allowed many species of forest and steppe mammals to pass from Eurasia to America (Kurtén, 1972).

During the last million years four major periods of intense cold (glacial periods) affected the northern hemisphere. Of these periods, the first and second each lasted 50 000 years. The second glacial period produced the most extensive ice-caps. Greenland, Iceland and the Faroe Islands were completely invested. In Europe ice covered Britain south to the Thames and

Severn rivers, and spread over Scandinavia, the Netherlands, Germany, the Alps, Poland and western USSR. Eastern Siberia had its own ice-caps, and North America was covered south of New York, St Louis and Vancouver. The third glacial period lasted 100 000 years, the fourth almost as long; both brought intense cold to the Arctic but their ice-caps were smaller.

The interglacials separating the cold periods included some spells of several thousands of years when temperate species of plants and animals were able to re-establish themselves in the Arctic. Within the glacial periods spells of intense cold (stadials) alternated with periods of warmth (interstadials), when temperate conditions were briefly restored. The final glacial period, which began 100 000 years ago, included three stadials or cold spells; during the second, warmer interstadial some 30 000 years ago, arctic summer temperatures rose above present levels, and many islands that are now tundra-covered were forested. During this period caribou, musk oxen, arctic hares, lemmings, arctic foxes and other polar species spread from Eurasia to America; moose, bison, sheep, lynx and other mammals of warmer climates also crossed Beringia, and all shifted south as colder conditions returned. The fourth glacial culminated some 20 000 years ago, to be followed by a slow, spasmodic warming that continues to the present day. Human history during this period is summarized briefly in Armstrong *et al.* (1978).

1.2.3 *The Subarctic*

The wide circumpolar belt of coniferous forest south of the treeline is dominated by spruce, pine, fir and larch, with broadleaf inclusions of alder, birch and poplar. Almost continuous across North America and northern Eurasia, it grows slowly on poor soils, in climates that, unlike those of temperate forests, would support very little else. Economic returns from this boreal forest are meagre, and except in the heavily populated areas of Europe and Canada, much of it remains almost untouched by man.

Boreal (meaning northern) forest is a good name for it. The Russian term taiga, meaning a marshy Siberian woodland, is a widely accepted alternative (Irving, 1972; Pruitt, 1978; Kimmins and Wein, 1986). However, some ecologists use taiga only for the transitional zone between tundra and forest, a zone for which the more descriptive name forest-tundra is preferable. Thus taiga is a confusing term, to be used with care or avoided altogether.

Köppen's subarctic boundary (see above) cuts through the boreal forest; only the northern half, including the forest-tundra, is climatically subpolar. By this definition subarctic lands include the central and western Aleutian islands, the Kommandorskiye Islands, much of central and southern Alaska, northern Canada, Labrador and Newfoundland, southern Iceland, central Scandinavia, and a wide swathe of European Russia, central Siberia and Kamchatka. As though to counter the asymmetry of the Arctic (see above) the subarctic zone is much wider in Siberia than in Canada. Short warm summers and long, often intensely cold winters characterize this zone. Permafrost is prevalent, and the zone includes the areas of central Canada and Siberia where the northern hemisphere's lowest winter temperatures are experienced (Chapter 2).

1.3 Antarctic and Subantarctic

1.3.1 *Antarctic topography*

Bounded by the Antarctic Convergence, the Antarctic Region includes continental Antarctica, Antarctic Peninsula, several groups of islands close by, and a band of ocean partly invested by sea ice. Antarctica is the fifth largest continent, with an area of about 14 million km^2, almost twice that of Australia. Its mean surface elevation is over 2000 m, more than twice that of Asia (960 m), the second highest. Without the ice-cap it would still be high, with a mean elevation of 850 m. Only about 2% of the continent is exposed; the rest lies under an ice mantle of mean thickness 1880 m (Drewry, 1983), that in places reaches over 4000 m thick (Figure 1.8). The ice forms a high central plateau rising to 4270 m, which falls gently at first and then steeply toward the edges, punctuated by nunataks (isolated mountain peaks). Most of the exposed rock appears along the Antarctic Peninsula, the Transantarctic Mountains, and on nunataks near the coast. About 10% of the coast is rocky; the rest is formed by steep ice cliffs, the edges of the covering mantle, that break away periodically to form massive tabular icebergs.

The continent is divided into East Antarctica, the main body, and West Antarctica which includes the peninsula and its roots; Greater and Lesser Antarctica are alternative names, now little used. East Antarctica is the mountainous bloc, formerly part of the supercontinent Gondwanaland, that drifted into juxtaposition with the South Pole during the mid-Tertiary, inaugurating the current ice age. Most of its mountains lie deeply buried under the ice. West Antarctica is a cluster of high volcanic islands, deeply dissected by channels and contiguous with the peninsula. Smaller than its neighbour, with a thinner ice-cap, it has many partly-exposed ranges including the Sentinel Range with Vinson Massif (5140 m), the continent's highest mountain. The huge bights of the Ross and Weddell Seas are backed

Figure 1.9 A polar environment; summer in McMurdo Sound, Antarctica. Glaciers sweep down from Mt Erebus, one of the continent's two active volcanos, to form ragged ice cliffs at sea level. Trails across the sea ice lead from McMurdo (left centre) and Scott Base (right).

by extensive ice plains, which rise inland to a saddle between the two major ice-caps.

The inland ice-cap is virtually sterile; its highest point is probably the world's coldest place. The South Pole, for over 30 years the site of Amundsen-Scott, a U.S. scientific station, is 1400 m lower and correspondingly warmer. Despite strong winds and aridity, rock surfaces in the exposed mountains, even within a few hundred kilometres of the South Pole, support tiny communities of algae, fungi, mosses, lichens, insects and mites (Chapter 3). The coast provides warmer and more congenial habitats, especially along the western flank of the Peninsula in summer, where mean monthly temperatures rise above freezing point.

The sea surrounding Antarctica is ice-encrusted for most of the year, fast ice growing close to the land each winter and joining a wide belt of pack ice that circles the continent endlessly. Several of the large archipelagos within the Antarctic Convergence, for example the South Shetland and South Orkney groups, are invested by sea ice for part or all of the year, and almost as cold as the continental shore. Other islands further north, for example South Georgia, lie beyond the pack ice and have much milder, more equable conditions throughout the year.

1.3.2 *Glacial and ecological history*

The fossil record of Antarctica, meagre because so little of the rock is exposed, testifies to a long pre-glacial history of warm or temperate climates. Coal measures were laid down close to the present South Pole, marine and terrestrial reptiles followed, and beech forests continued to flourish long after the first glaciers formed. Ice-caps have covered the continent to about their present extent for 2 million years, possibly longer, and their huge masses have responded only slightly to the glacial cycles that so dramatically affected the north. None of the present continental flora and fauna is known to be relict from pre-glacial times. Islands close to the continent show similar glacial histories, mostly indicating lower temperatures and more widespread glaciation in the recent past.

Antarctica was formerly linked to South America and Australia in the Gondwana supercontinent, and no doubt shared a Tertiary flora and fauna with them. Since its separation some 20 million years ago the continent and its nearby islands have been isolated from the rest of the world. Though a few island species may have survived in refugia, the post-glacial biota of most of the area is restricted to creatures that have drifted in downwind, swum to their shores or been carried by birds. Man is a late arrival; though many of the subantarctic islands were known to sealers in the eighteenth and early nineteenth centuries, he first saw mainland Antarctica in 1820 and probably first set foot on it in 1895 (Mill, 1905; *Reader's Digest*, 1985).

1.3.3 *The Subantarctic*

Between the Antarctic and Subtropical Convergences in the southern Atlantic, Indian and Pacific Oceans lie 11 scattered groups of islands collectively called the subantarctic islands; the zone includes also New Zealand's South Island and the tip of South America. The islands range latitudinally from Macquarie Island in 54°S to Iles Amsterdam in 37°S; not surprisingly they have a wide range of climates, soil conditions and ecology. Many are volcanic and of geologically recent origin; some are probably no older than the antarctic ice-cap. Though all are colder than they would be in equivalent northern latitudes, a better collective name is the southern temperate islands (Stonehouse, 1982). The warmest lie north of the 10°C summer isotherm and support shrubs of tree height; the southernmost are cold, wet and gloomy, constantly buffeted by westerly winds, but thickly covered with soil, grasses and forbs.

Only the largest, the main island of Iles Kerguelen, standing close to the Antarctic Convergence, carries remnants of an ice-cap; its mountain glaciers show ample evidence of recent shrinkage, though several still reach the sea. Iles Crozet and other southern islands are snow-covered in winter, and often carry snow in summer after the passage of depressions. The smaller islands may never have been ice-capped; their climatic regimes, currently among the most equable and constant in the world, have probably altered little since the islands rose from the sea.

1.4 Summary and conclusions

This chapter defines the polar regions, the boundaries ascribed to them (10°C summer isotherm, treeline, maritime discontinuities), and the special qualities that make them ecologically different from the rest of the world; only alpine regions are comparable, and in only a few respects. The geography of the two regions—the mostly low-lying arctic shores surrounding a deep ocean basin, and the high antarctic continent surrounded by moat-like ocean—is described with particular reference to glacial history, conditions for life, accessibility to invading species, and other factors affecting their present-day ecology. The subpolar fringes—in the north the cold continental heartlands, in the south the subantarctic oceanic islands—are also described.

CHAPTER TWO
ENERGY, CLIMATE AND MICROCLIMATES

2.1 Introduction: why polar cold?

The climatic elements that most concern living systems the world over are ground temperature, air temperature, wind, humidity and precipitation. All depend ultimately on the amount of solar radiation received. Polar regions are cold because they receive their radiation obliquely and reflect much of it away rather than absorbing it. A column of sunlight 1 m square, shining on a tropical region, passes through a thin stratum of atmosphere and warms of ground beneath (Figure 2.1). A similar column shining on a polar region passes through a greater thickness of atmosphere, where more of the energy is reflected or absorbed, and spreads tangentially over a larger area; then it is mostly reflected away by snow and ice surfaces. It cannot warm the earth to the same degree, and the earth cannot warm the overlying atmosphere. In consequence everything feels colder.

Solar radiation becomes ineffective for warming at angles of incidence below about 17°. Polar regions receive annually only about 40% as much

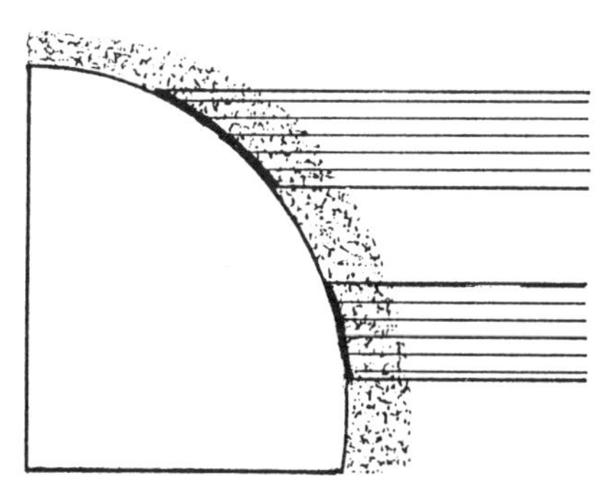

Figure 2.1 Solar rays striking high latitudes pass obliquely through the atmosphere, and spread over a wider area than those illuminating lower latitudes.

solar radiation as equatorial regions, and reflect away on average some 89–90% of all they receive. Most of their energy they receive, reflect and re-radiate during the long summer days (Table 2.1); though the sun's angle never exceeds about 45°, daily totals of radiation received under clear summer skies are among the world's highest. Being warmer than space even in winter, they continue to radiate energy through the sunless period when incoming radiation is minimal, incurring a strong energy deficit. For the year as a whole the Arctic has a slight positive radiation balance, the Antarctic a marked deficit. Their mean temperatures are maintained by heat brought in from lower latitudes by the atmosphere and oceans. Antarctica currently receives about 7% more radiation than the Arctic, because southern summer occurs when the earth is at perihelion, or closest to the sun.

With all its shortcomings, polar sunshine packs significant radiant energy; like winter sunshine on clear days in temperate latitudes it is strong enough to feel warm, to warm the environment gently for plants and animals and to stimulate photosynthesis. Domestic solar heating panels can be effective in summer even close to the Arctic Circle (MacGregor, 1985). Spring thaw in both polar regions is initiated by surface warming from direct insolation, and the autumn freeze-up starts as daily insolation weakens. Because of their direct bearing on microclimates, the small-scale climatic environments in which organisms actually live, insolation and radiation balance are particularly important factors in polar ecology (Figure 2.2).

2.2 Solar energy

Despite its molten interior and thin crust, the earth receives less than 0.1% of its surface heat from within; virtually all the heat of the ground and atmosphere comes from the sun. The range of wavelengths emitted by a radiating body is an inverse function of the body's absolute temperature (in K, i.e. temperature in °C plus 273). With a mean surface temperature of about 5800 K the sun radiates mostly short waves of length 0.1 μm to 5.0 μm, including ultraviolet (UV). (0.1–0.4 μm), visible light (0.4–0.7 μm) and infrared (IR) (0.7–5.0 μm). Radiation is monitored using instruments (pyrometers, pyronometers, radiometers) that measure differential warming effects on black (absorbent) and silvered (reflective) surfaces. Filters select the spectrum of rays to be metered, and the instruments can be mounted facing upward or downward, on the ground, or on aircraft and satellites, to measure radiation at any level and from any direction.

Table 2.1 Hours of daylight on the 21st of each month, and dates of start and end of 24-hour periods of light and darkness, in latitudes 50° to 90° north and south. These times and dates are correct for 1989, and vary slightly from year to year. Data calculated by Dr Gareth Rees.

		Latitude south					Latitude north				
		90°	80°	70°	60°	50°	50°	60°	70°	80°	90°
January		24.0	24.0	24.0	17.5	15.6	8.8	7.1	2.5	0.0	0.0
February		24.0	24.0	16.5	14.7	13.9	10.5	9.8	8.3	2.3	0.0
March		24.0	12.4	12.2	12.1	12.1	12.2	12.3	12.5	13.6	24.0
April		0.0	0.0	7.7	9.4	10.2	14.1	15.1	17.1	24.0	24.0
May		0.0	0.0	2.0	7.0	8.7	15.7	17.6	24.0	24.0	24.0
June		0.0	0.0	0.0	5.9	8.0	16.4	18.9	24.0	24.0	24.0
July		0.0	0.0	1.6	7.0	8.7	15.7	17.7	24.0	24.0	24.0
August		0.0	0.0	7.6	9.4	10.2	14.1	15.1	17.2	24.0	24.0
September		24.0	12.2	12.1	12.1	12.1	12.3	12.4	12.6	13.1	24.0
October		24.0	24.0	16.6	14.8	13.9	10.4	9.7	8.2	0.8	0.0
November		24.0	24.0	24.0	17.5	15.6	8.8	7.1	2.4	0.0	0.0
December		24.0	24.0	24.0	18.9	16.4	8.1	5.9	0.0	0.0	0.0
24-h day	starts	21/9	17/10	18/11	—	—	—	—	16/5	13/4	19/3
	ends	22/3	24/2	21/1	—	—	—	—	27/7	29/8	24/9
24-h night	starts	23/3	18/4	25/5	—	—	—	—	26/11	22/10	25/9
	ends	20/9	26/8	18/7	—	—	—	—	16/1	19/2	18/3

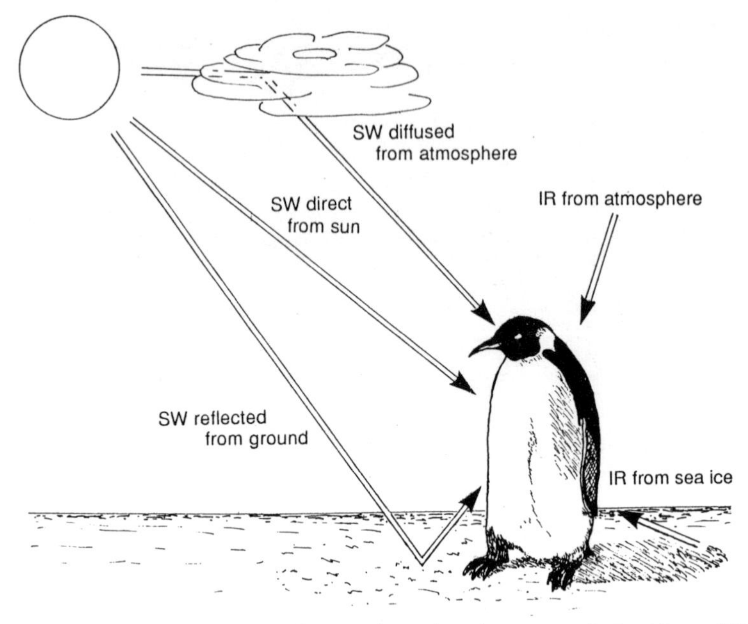

Figure 2.2 Radiation environment. The penguin receives shortwave radiation (direct, diffuse and reflected) from the sun, and longwave radiation from the atmosphere and substrate.

2.2.1 *Energy budgeting*

The outer layer of atmosphere facing the sun receives radiation at a mean rate of 1.98–2.0 cal m^{-2} min^{-1}. Some of this energy is reflected immediately back to space; the rest passes into the atmosphere and through it to earth. Both atmosphere and earth are reflectors, so that altogether some 63% of incoming radiation returns to space unabsorbed. The remaining 37% is absorbed by earth and atmosphere, warming both, and both then radiate energy themselves, at longer wavelengths. Overall some 10–20% of the earth's emissions are caught by the atmosphere, which is further warmed by conduction and convection from the earth. About two-thirds of the atmosphere's warmth comes from earth, the rest directly from the sun. Most of these transfers occur in tropical and subtropical regions, where insolation is strongest.

How that energy is distriubuted in polar regions is shown for selected stations in Tables 2.2 and 2.3; localities of the stations are given in Figures 2.3 and 2.4. Of the antarctic stations, Vostok is on the high plateau, Mirny is coastal in an area of relatively high snow-cover, and Oasis is a

Table 2.2 Monthly and yearly radiation budgets for Arctic polar stations Dikson, Mys Chelyuskin and Mys Schmidt. All values are in kcal cm^{-2}. For station localities see Table 2.4 and Figure 2.3. Data after Gavrilova (1963).

Dikson													
Month	Jan	Feb	Mar	Apr	May	Jun	Jul	Aug	Sep	Oct	Nov	Dec	Year
DirSWR	0.0	0.0	1.8	3.7	3.2	3.7	4.2	2.0	0.8	0.2	0.0	0.0	19.6
DifSWR	0.0	0.4	2.4	6.1	11.1	10.1	7.7	5.0	2.4	0.8	0.0	0.0	46.0
TSWR	0.0	0.4	4.2	9.8	14.3	13.8	11.9	7.0	3.2	1.0	0.0	0.0	65.6
RSWR(−)	0.0	0.1	0.6	1.5	2.1	8.1	9.5	5.1	2.1	0.2	0.0	0.0	29.3
ESWR	0.0	0.3	3.6	8.3	12.2	5.7	2.4	1.9	1.1	0.8	0.0	0.0	36.3
ELWR(−)	2.4	2.1	2.6	2.2	1.8	1.9	1.9	1.9	1.9	2.3	2.5	2.3	25.8
NRB	−2.4	−1.8	1.0	6.1	10.4	3.8	0.5	0.0	−0.8	−1.5	−2.5	−2.3	10.5
Mys Chelyuskin													
Month	Jan	Feb	Mar	Apr	May	Jun	Jul	Aug	Sep	Oct	Nov	Dec	Year
DirSWR	0.0	0.0	1.1	3.5	3.9	3.8	4.4	1.1	0.5	0.0	0.0	0.0	18.3
DifSWR	0.0	0.1	1.6	5.5	11.3	12.1	7.6	5.0	1.9	0.4	0.0	0.0	45.5
TSWR	0.0	0.1	2.7	9.0	15.2	15.9	12.0	6.1	2.4	0.4	0.0	0.0	63.8
RSWR(−)	0.0	0.0	0.5	1.9	3.5	6.7	9.4	5.0	1.9	0.2	0.0	0.0	29.1
ESWR	0.0	0.1	2.2	7.1	11.7	9.2	2.6	1.1	0.5	0.2	0.0	0.0	34.7
ELWR(−)	2.4	2.1	2.6	2.6	2.3	2.5	2.0	2.0	1.6	2.1	2.0	2.6	26.8
NRB	−2.4	−2.0	−0.4	4.5	9.4	6.7	0.6	−0.9	−1.1	−1.9	−2.0	−2.6	7.9
Mys Schmidt													
Month	Jan	Feb	Mar	Apr	May	Jun	Jul	Aug	Sep	Oct	Nov	Dec	Year
DirSWR	0.0	0.3	2.2	4.6	4.6	5.8	4.7	2.6	1.6	0.4	0.0	0.0	26.8
DifSWR	0.0	0.8	3.1	5.9	10.2	9.0	6.9	4.9	2.6	0.8	0.2	0.0	44.4
TSWR	0.0	1.1	5.3	10.5	14.8	14.8	11.6	7.5	4.2	1.2	0.2	0.0	71.2
RSWR(−)	0.0	0.2	1.1	2.1	3.0	8.4	10.5	6.0	2.4	0.9	0.0	0.0	34.6
ESWR	0.0	0.9	4.2	8.4	11.8	6.4	1.1	1.5	1.8	0.3	0.2	0.0	36.6
ELWR(−)	2.8	2.1	3.0	2.6	2.1	2.5	2.2	2.1	2.5	2.4	2.1	2.1	28.5
NRB	−2.8	−1.2	1.2	5.8	9.7	3.9	−1.1	−0.6	−0.7	−2.1	−1.9	−2.1	8.1

Table 2.3 Monthly and yearly radiation budgets for Antarctic polar stations Mirny, Oasis and Vostok. All values are kcal cm^{-2}. Note that, to facilitate summer and winter comparisons with Arctic stations in Table 2.2, the columns are tabulated from July onward. For station localities see Table 2.5 and Fig. 2.4. Data after Rusin (1964).

Mirny													
Month	Jul	Aug	Sep	Oct	Nov	Dec	Jan	Feb	Mar	Apr	May	Jun	Year
DirSWR	0.0	0.8	2.2	5.7	10.3	12.8	10.6	7.5	4.0	0.9	0.1	0.0	54.9
DifSWR	0.1	0.7	2.9	5.3	7.6	9.1	9.1	6.4	3.6	1.7	0.2	0.0	46.7
TSWR	0.1	1.5	5.1	11.0	17.9	21.9	19.7	13.9	7.6	2.6	0.3	0.0	101.6
RSWR(−)	0.1	1.4	4.6	9.3	15.0	17.7	15.8	11.7	6.6	2.2	0.3	0.0	84.7
ESWR	0.0	0.1	0.5	1.7	2.9	4.2	3.9	2.2	1.0	0.4	0.0	0.0	16.9
GLWR(−)	0.7	10.5	11.5	11.5	12.6	13.8	13.8	12.7	12.0	11.5	11.1	10.8	142.5
ALWR	8.8	8.6	9.5	9.5	11.2	12.3	11.9	11.7	9.6	9.6	9.4	8.5	120.6
ELWR(−)	1.9	1.9	2.0	2.0	1.4	1.5	1.9	1.0	2.4	1.9	1.7	2.3	21.9
NRB	−1.9	−1.8	−1.5	−0.3	1.5	2.7	2.0	1.2	−1.4	−1.5	−1.7	−2.3	−5.0
Oasis													
Month	Jul	Aug	Sep	Oct	Nov	Dec	Jan	Feb	Mar	Apr	May	Jun	Year
DirSWR	0.0	0.6	2.7	4.8	7.3	9.2	8.7	6.2	3.0	1.1	0.4	0.0	44.0
DifSWR	0.0	0.7	2.8	4.7	8.0	9.5	8.4	5.8	3.6	1.5	0.2	0.0	45.2
TSWR	0.0	1.3	5.5	9.5	15.3	18.7	17.1	12.0	6.6	2.6	0.6	0.0	89.2
RSWR(−)	0.0	0.5	2.3	2.5	3.1	3.5	2.7	2.6	2.1	0.9	0.3	0.0	20.5
ESWR	0.0	0.8	3.2	7.0	12.2	15.2	14.4	9.4	4.5	1.7	0.3	0.0	68.7
GLWR(−)	9.8	9.9	10.3	11.4	13.3	14.7	14.9	13.7	12.3	11.6	10.7	9.0	141.6
ALWR	8.0	8.0	8.1	11.3	8.6	11.4	11.0	11.1	10.5	8.5	8.6	7.0	112.1
ELWR(−)	1.8	1.9	2.2	0.1	4.7	3.3	3.9	2.6	1.8	3.2	2.1	2.0	29.5
NRB	−1.8	−1.1	1.0	6.9	7.5	11.9	10.5	6.8	2.7	−1.4	−1.8	−2.0	39.2
Vostok													
Month	Jul	Aug	Sep	Oct	Nov	Dec	Jan	Feb	Mar	Apr	May	Jun	Year
DirSWR	0.0	0.0	1.8	8.7	20.0	25.6	22.3	14.8	4.2	0.4	0.0	0.0	97.8
DifSWR	0.0	0.0	0.8	3.3	4.1	4.7	5.8	2.0	1.6	0.0	0.0	0.0	22.3
TSWR	0.0	0.0	2.6	12.0	24.1	30.3	28.1	16.8	5.8	0.4	0.0	0.0	120.1
RSWR(−)	0.0	0.0	2.5	10.3	20.0	24.6	23.1	13.8	5.1	0.3	0.0	0.0	99.7
ESWR	0.0	0.0	0.1	1.7	4.1	5.7	5.0	3.0	0.7	0.1	0.0	0.0	20.4
GLWR(−)	4.5	4.0	4.4	5.2	7.0	8.5	8.8	6.8	5.3	4.6	4.7	4.3	68.1
ALWR	3.5	2.8	3.4	3.2	3.3	3.6	4.4	3.7	3.7	3.4	4.1	3.5	42.5
ELWR(−)	1.0	1.2	1.0	2.0	3.7	4.9	4.4	3.1	1.7	1.2	0.6	0.8	25.6

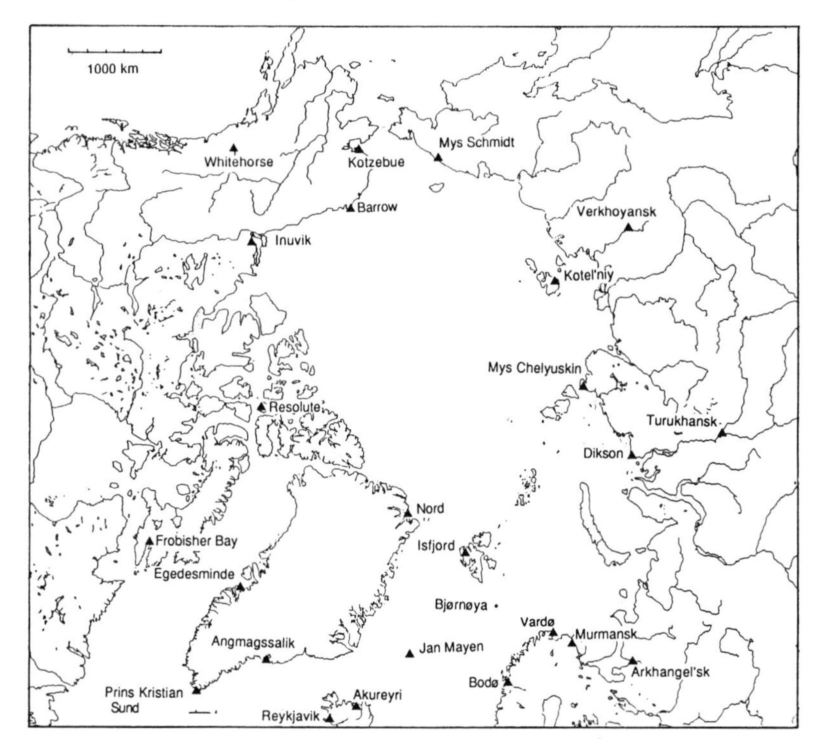

Figure 2.3 Localities of Arctic stations listed in Tables 2.2 and 2.4.

bare rocky area near Mirny (Chapter 3). The Arctic has no high-altitude, high-latitude station equivalent to Vostok, but several stations close to sea level in high northern latitudes have radiation regimes similar to Mirny and Oasis.

Over the world as a whole some 25–30% of the sun's incident radiation is reflected back into space by atmospheric particles—cloud, water vapour and dust. Over the cloudy subpolar regions more than average is reflected back. Nearer the poles the percentage is less, for skies are clearer, clouds are thinner, and the air is usually clean and dry. The atmosphere absorbs about 20% of the radiation, mostly UV which is taken up by oxygen and ozone molecules. In polar regions and at high altitudes, where the atmosphere is thin, the UV radiation passing through is intense enough to cause rapid sun tanning and severe conjunctivitis (snow-blindness) in humans exposed to it. Radiation of longer wavelengths is absorbed selectively by water vapour

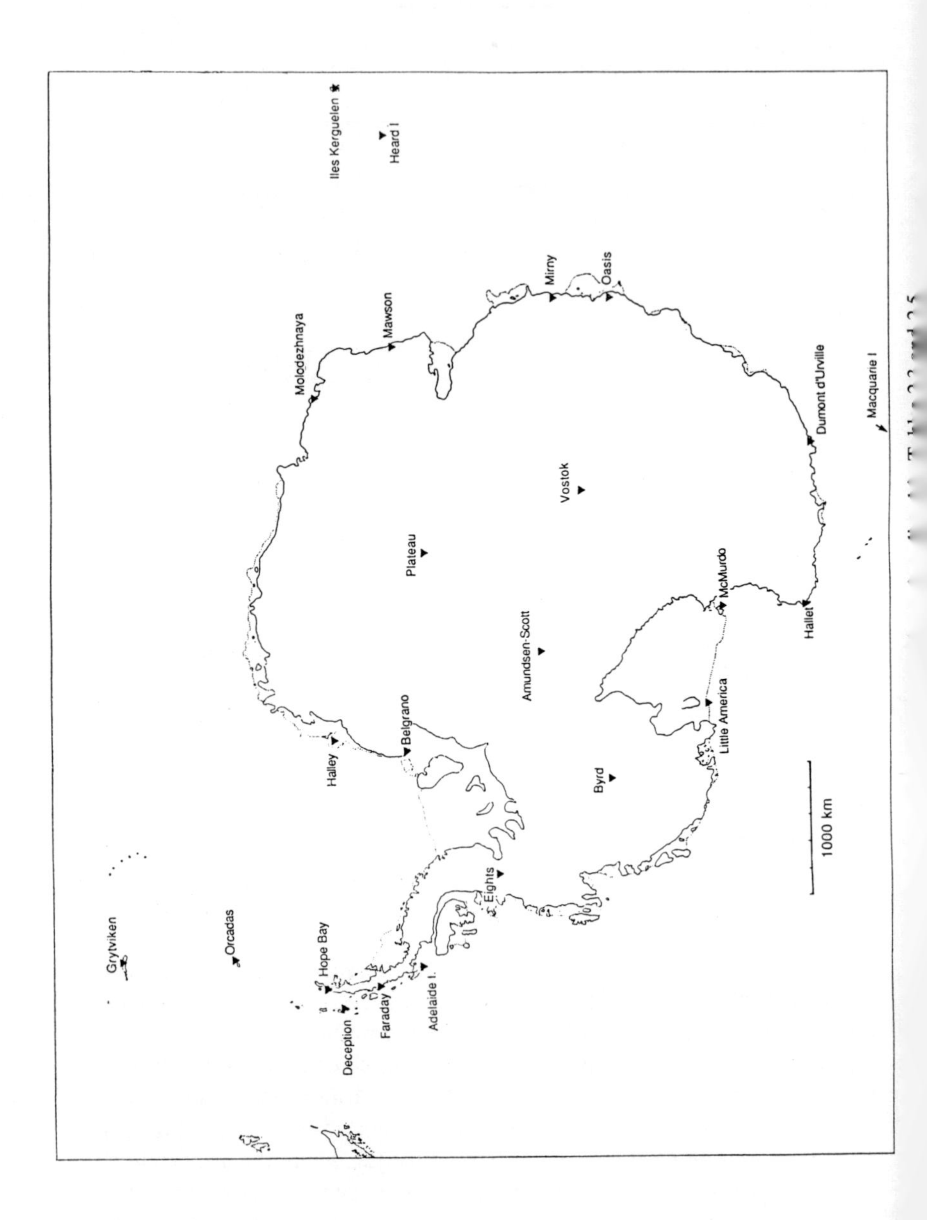
Iles Kerguelen
Heard I
Molodezhnaya
Mawson
Mirny
Oasis
Dumont d'Urville
Macquarie I
Vostok
Plateau
McMurdo
Hallet
Amundsen-Scott
Belgrano
Halley
Little America
Byrd
1000 km
Eights
Grytviken
Orcadas
Hope Bay
Deception
Faraday
Adelaide I.

and carbon dioxide; cloud absorbs impartially from almost the whole spectrum. However, most of the energy received on the ground is short-wave radiation.

Concern is currently being expressed that a depletion of ozone, recently detected above Antarctica (the ozone hole), and probably occurring above both polar regions, is altering radiation balance by allowing more UV through to the earth's surface during polar summers. The depletion may be due mainly to chlorofluorocarbons and other man-made industrial gases, which escape to the atmosphere and break down in the upper atmosphere to release radicals which destroy ozone. These gases are currently entering the atmosphere five times faster than natural processes are destroying them, with results which may be environmentally harmful (Farman *et al.* 1985; Farman, 1987).

Short-wave input. The radiation that reaches the earth has been diminished by filtration and also divided. Part comes directly from the sun (direct short-wave radiation, DirSWR); the rest is diffused through clouds and other atmospheric particles (diffuse short-wave radiation, DifSWR). Together these make up total short-wave radiation (TSWR). As Table 2.3 shows, direct radiation far exceeds diffuse radiation through the clear atmosphere over Vostok. Near sea level at Oasis and Mirny the two components are roughly aquatic. In the Arctic basin diffuse radiation through thin stratus cloud nearly always exceeds direct radiation.

Annual values of TSWR, measured in kcal cm^{-2}, range from over 200 in sunny tropical uplands to 120 in mid-latitude and 80 in the overcast fifties and sixties. In northern polar regions values of 60 to 80 are common, being highest over dry northern Greenland and lowest over the nearby Norwegian Sea, where warm maritime air brings heavy clouds (Vowinckel and Orvig, 1970). In the far south they are higher despite the sunless winter months. The high antarctic plateau has among the world's highest annual TSWR, in places exceeding 300 kcal cm^{-2}.

Albedo. That the same region has the world's coldest climates is a tribute to its high reflectivity and the clarity of the atmosphere above. Reflected short-wave radiation (RSWR) is the amount of short-wave radiation reflected back into space. Overall 70–90% is reflected from polar regions; some is re-reflected downward from clouds or absorbed into the cloud cover, but the rest leaves the earth–atmosphere system altogether. The ratio RSWR/TSWR (often multiplied by 100 and expressed as a percentage) is the reflectivity or albedo (whiteness) of the surface. Figure 2.5 gives typical

Surface	albedo
Fresh snow, snow-covered sea ice, white clouds, white-capped seas	75-95
Older snow, new (grey) sea ice, rough seas, pale sand	40-75
Smooth pale rocks	20-40
Rough dark rocks, calm seas	5-20

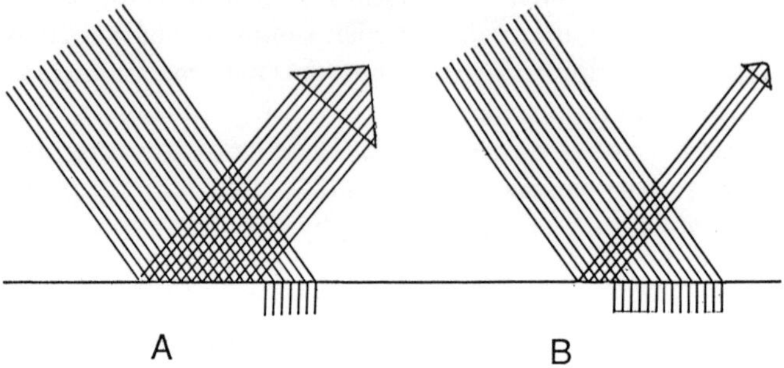

A B

Figure 2.5 Albedo. With high albedo (A, for example Mirny in Table 2.3) much of the incident radiation is reflected away, and only a little remains to warm the ground. With low albedo (B, for example Oasis), less is reflected and more absorbed; a lower total incidence can result in higher ground temperatures.

polar albedo values; others may be calculated from mean monthly and annual values of TSWR and RSWR in Tables 2.2 and 2.3.

Effective short-wave radiation. RSWR subtracted from TSWR leaves effective short-wave radiation (ESWR), the amount available to penetrate soil, rock, snow and water and yield up its energy to heat the surface. At Vostok year-round snow gives mean monthly albedos of about 80%; only 16.9% of the year's TSWR is absorbed as ESWR. By contrast at Oasis mean monthly albedos range from 16% in summer to 50% in winter, and mean annual albedo is 23%. 77% of the year's short-wave radiation is effective. Though Oasis receives less short-wave energy per year than the other antarctic stations, it absorbs three times as much as Vostok and four times as much as neighbouring Mirny; not surprisingly its surface temperatures are generally much higher. The arctic stations show a wide range of albedos throughout the year depending mainly on seasonal variation in snow-cover, with annual values around 45–50%.

Long-wave output. Energized by short-wave radiation, the earth's surface itself radiates energy. With a mean surface temperature of only 285 K its emmisions are mostly long waves of between 4.5 and 50 μm with a peak of energy production at about 10 μm. The amount of energy radiated is a function of the IR emissivity of the surface and the fourth power of its absolute temperature. Perfect IR emissivity (called black body radiation) in fact relates more to surface texture than to colour. For most natural surfaces emissivities range between 80% and approaching 100%. Snow, water, ice and rock at the same temperature all have similar high emissivities.

The ground long-wave radiation (GLWR), derived from absorbed short-wave radiation, is partly absorbed by the atmosphere, notably by carbon dioxide, ozone, water vapour and cloud, and the atmosphere is warmed by the energy it absorbs. The atmosphere in turn transmits atmospheric long-wave radiation (ALWR), about 75% of which returns and reduces the earth's net heat loss. The remaining IR from both sources (mostly in the spectral range 8.5–11 μm, to which water vapour is virtually transparent) passes through the atmosphere and is lost to space. In radiation budgets GLWR is negative and ALWR positive. Their sum, effective long-wave radiation (ELWR), is negative for earth and atmosphere as a whole, though it is sometimes positive locally, for example when a layer of warm cloud moves in to overlie cold ground. This is an important source of heat at polar stations when depressions move in from lower latitudes. GLWR and ALWR (here given only for the antarctic stations) can be measured directly but are usually calculated. GLWR from the absolute temperature of the surface, and ALWR from temperature and humidity parameters in the atmosphere and troposphere.

Radiation balance. Net radiation balance (NRB), the sum of ESWR and ELWR, is distributed world-wide as shown in Figure 2.6. For most polar areas NRB is positive in summer and negative in winter, losses equalling or exceeding gains over the year as a whole. On the south polar plateau monthly values are positive only in high summer and there is a marked annual deficit; Vostok has negative values for nine months. Mirny for eight months, and their annual losses are respectively 5.2 and 5.0 kcal cm^{-2}. Similar values have been obtained at other coastal stations where snow is present for much of the year. Snow-free Oasis shows positive values for seven months and a high positive annual balance. At Scott Base, a coastal station in area of low snowfall further south than Oasis, a positive annual balance of over 18 kcal cm^{2-} has been recorded (Thompson and Macdonald, 1962). In Wright Valley, a nearby oasis area with a longer snow-free season and less cloud, Bull (1966) estimated a net annual gain of 29 kcal cm^{-2}.

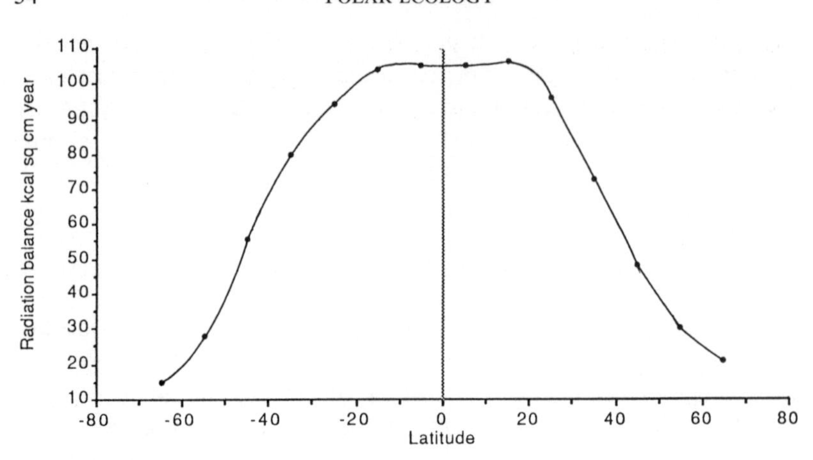

Figure 2.6 Radiation balance for latitudes between 80°S and 80°N. Data from Budyko (1977), augmented.

In the Arctic most coastal stations have positive values of NRB for four to six months each summer. Annual values range from slightly negative over much of the polar basin to slightly positive over coastal areas, reaching 10–15 kcal cm^{-2} along the Arctic Circle (Gavrilova, 1963). Between 50° and 60° annual balances of 20–40 kcal cm^{-2} prevail in both hemispheres, and are typical of many northern population centres. Clearly, similar values of NRB would be found in both polar regions but for the relatively high albedos of the ice-cover. It is not difficult to see how, in a world without polar ice, temperate conditions could extend far into polar regions, and temperate vegetation flourish well beyond present polar limits.

Greenhouse effect. The atmosphere forms a protective greenhouse about the earth, letting in short-wave energy but inhibiting the escape of the long-wave energy derived from it. Without the atmosphere the earth's surface temperature would average 30–40°C cooler, and vary far more diurnally and seasonally, like the surface of the moon.

The greenhouse effect is strong in the Arctic, particularly on the periphery where low cloud is prevalent. The antarctic atmosphere, generally clearer, drier, and thinner because of the high plateau, is more transparent to IR radiation and less effective as a greenhouse. In both hemispheres temperature inversions (local increases in temperature with height) enhance the effect. They occur in still air or light winds, when radiation from the ground (usually a snow surface) is intense, and heat flow from atmosphere to ground is inhibited. Once formed, the inversion reduces outgoing radiation and keeps the surface warmer than it would otherwise be. Inversions exist almost permanently over the antarctic plateau in all but the warmest months, and form rapidly at coastal stations during calm, anticyclonic conditions. Their persistence helps to explain why temperatures on the antarctic plateau, after a rapid fall in autumn, continue to fall only slightly after mid-winter (see below, Table 2.5 and Figure 2.11).

2.2.2 *Heat influx to polar regions*

The annual drain of radiant energy from polar regions is balanced by inflows of heat from lower latitudes. The Arctic has three major sources of influx. Warm ocean currents, notably the North Atlantic Drift, carry warmth from the Gulf of Mexico to the shores of Greenland, Iceland, Svalbard and northern Eurasia. Heat pours in too from the great Siberian rivers, which flow northward from the Asian heartland. Most important of all, air masses with a northern component contribute both sensible heat (i.e. direct warmth) and latent heat in their water vapour.

The Antarctic gains very little from ocean currents, which tend to flow peripherally about the continent, and nothing from rivers; almost all of its heat influx comes from air masses, which constantly circle the continental coast and occasionally whirl inland toward its centre. In either hemisphere it is the circulating air masses that bring most warmth, and indeed most of the day-to-day weather to polar regions, and ultimately determine their climates.

The atmosphere absorbs heat mainly from the earth, by radiation and convection in passing over warm surfaces, and by absorbing water vapour from warm water or warm moist ground. Water vapour's bonus of latent heat (539 cal g^{-1}), acquired in its transformation from liquid to vapour, is released when the vapour condenses, and more (79.8 cal g^{-1}) is released when the resulting water freezes in the upper atmosphere.

Turbulent exchanges. Heat is transferred from warm atmosphere to cold ground by turbulent heat exchange (L) and turbulent moisture exchange (V), values of which can be calculated for each month from recorded wind velocities, temperatures, humidity values and other parameters. Their sum for the year should tally with annual values of B, the net radiation balance, but it is not surprising to find that no single station yields an exact balance (Rusin, 1964; Liljequist, 1956). At antarctic plateau stations where V is negligible, L and B are approximately balanced; on the steep slopes of the Antarctic continent V is small and L slightly higher than on the plateau. At points on the coast of Antarctica where katabatic (downslope) winds are strong, both *L* and *V* are high, and their sum approximates to local values of *B*. In coastal areas where winds are slight *V* is again low, and *L* and *B* become complementary.

2.2.3 *Radiation and microclimate*

Weather data recorded by meteorologists are collected under standard conditions that facilitate comparisons between stations. Air temperatures and humidities are measured in louvred boxes or screens at 2 m above ground level that protect the instruments from direct sunlight, wind direction and strength are measured on masts 10 m above the ground, and precipitation is collected in gauges designed to minimize evaporation losses.

Few organisms live in the conditions represented by these measurements, indeed most seek the antitheses; polar organisms particularly relish direct sunlight, high humidity and shelter from wind. Standard climatic summaries for polar stations (Tables 2.4, 2.5) are thus useful in their main purpose of comparing one station with another, but seldom good guides to the actual living conditions of organisms. For these we need microclimate data, measured in soil, among rocks or under vegetation, as close as possible to the organisms themselves. For a review of the microclimate concept and methods of measuring see Geiger (1965); Walton (1982a) discusses techniques applicable in polar regions.

While many microclimate studies have been made at polar sites, particularly in the past decade, most have used inadequate or suspect techniques; Walton (1984) comments critically on over a dozen antarctic studies. However, results generally show that differences between micro- and macroclimate are considerable, and microclimate should wherever possible be taken into account when living conditions of terrestrial organisms are being assessed.

Climate near the ground. Rapid but slight changes in solar radiation and wind at ground level, which escape detection by normal meteorological recording, cause striking shifts in the microclimates of plants and animals. Their moment-to-moment effects are perhaps seen most clearly in spring, when air temperature are below but close to freezing point. At such times rock and snow surfaces take their temperature from the air but are readily warmed by shafts of sunlight—direct short-wave radiation—moving across them. With an albedo of about 10% (Figure 2.5) the rocks absorb much of the energy and their surface temperature rises. Rock conducts heat

Table 2.4 Locations, elevations and mean monthly and annual temperatures at selected Arctic stations, Data from WMO (1971).

Station	Position	height (m)	Jan	Feb	Mar	Apr	May	Jun	Jul	Aug	Sep	Oct	Nov	Dec	Mean	Range
North America																
Barrow	71°18′N 156°47′W	4	−26.8	−27.9	−25.9	−17.7	−7.6	0.6	3.9	3.3	−0.8	−8.6	−18.2	−24.0	−12.4	31.8
Kotzebue	66°52′N 162°38′W	5	−20.9	−20.0	−18.9	−10.4	−0.6	6.6	11.5	10.3	4.9	−4.1	−13.7	−19.8	−6.3	32.4
Whitehorse	60°43′N 135°04′W	698	−18.1	−14.1	−7.6	0.1	7.5	12.6	14.2	12.4	7.9	0.7	−8.2	−15.1	−0.7	32.3
Inuvik	68°18′N 133°39′W	61	−28.8	−27.0	−22.6	−12.7	−0.7	9.4	13.6	10.4	3.4	−7.0	−19.9	−27.0	−9.1	32.2
Resolute	74°43′N 94°59′W	64	−31.8	−33.7	−31.4	−22.1	−10.2	0.6	4.6	2.9	−4.4	−14.6	−24.9	−29.3	−16.2	38.3
Frobisher Bay	63°45′N 68°33′W	21	−26.1	−24.8	−21.2	−13.2	−2.2	3.9	8.1	7.0	2.7	−5.0	−13.8	−21.7	−8.8	34.2
Greenland, Iceland																
Egedesminde	68°42′N 52°52′W	47	−14.2	−15.8	−15.7	−9.3	−0.7	3.7	6.4	6.1	2.7	−1.8	−5.9	−10.8	−4.6	22.2
Prins Kristian Sund	60°03′N 43°12′W	76	−4.4	−3.8	−2.8	−0.5	2.3	4.8	7.0	7.4	5.4	2.4	−0.5	−3.4	1.2	11.8
Angmagssalik	63°37′N 37°39′W	35	−7.5	−7.8	−6.4	−3.5	1.4	4.9	6.6	6.6	4.1	0.3	−2.8	−5.7	−0.8	14.4
Nord	81°36′N 16°40′W	35	−28.7	−30.4	−31.0	−24.0	−11.2	−0.4	3.9	1.4	−8.5	−18.3	−23.8	−27.3	−16.5	34.9
Reykjavik	64°08′N 21°56′W	16	−0.4	−0.1	1.5	3.1	6.9	9.5	11.2	10.8	8.6	4.9	2.6	0.9	5.0	11.6
Akureyri	65°41′N 18°05′W	5	−1.5	−1.6	−0.3	1.7	6.3	9.3	10.9	10.3	7.8	3.6	1.3	−0.5	3.9	12.5
Svalbard, northern Norway																
Jan Mayen	70°57′N 08°40′W	9	−4.0	−5.2	−4.8	−3.4	−0.5	2.4	5.2	5.5	3.8	0.9	−1.2	−2.9	−0.1	10.7
Isfjord	78°04′N 13°38′E	9	−10.3	−9.9	−11.9	−8.2	−2.7	2.1	5.0	4.5	1.3	−2.4	−5.3	−7.9	−3.8	16.9
Bjørnøya	74°31′N 19°10′E	14	−5.3	−6.2	−7.0	−5.2	−0.8	2.4	4.5	5.2	3.0	0.6	−1.6	−3.6	−1.2	12.2
Vardø	70°22′N 31°06′E	15	−4.3	−5.2	−4.0	−0.8	2.6	6.2	9.1	9.7	6.8	2.5	−0.5	−2.7	1.6	14.9
Bodø	67°17′N 14°25′E	13	−2.1	−2.4	−1.0	2.2	6.2	9.9	13.6	12.7	9.4	5.1	1.9	−0.1	4.6	16.0
USSR																
Murmansk	68°58′N 33°03′E	46	−10.9	−11.4	−8.1	−1.4	3.9	10.0	13.4	11.1	6.9	0.9	−3.8	−7.9	0.2	24.8
Arkhangel'sk	64°35′N 40°30′E	13	−11.7	−11.7	−8.1	−0.1	5.9	13.0	16.3	14.5	8.3	1.9	−3.4	−8.6	1.4	28.0
Dikson	73°30′N 80°14′E	20	−24.1	−24.3	−25.0	−16.4	−7.0	0.7	4.9	5.4	1.9	−6.6	−17.0	−21.4	−10.7	30.4
Turukhansk	65°47′N 87°57′E	32	−25.6	−22.8	−16.9	−7.2	0.8	10.8	16.1	12.6	6.3	−4.5	−19.7	−24.9	−6.2	41.7
Mys Chelyuskin	77°43′N 104°17′E	13	−27.3	−27.4	−28.0	−21.0	−9.7	−1.0	1.5	0.8	−2.2	−10.3	−20.1	−25.0	−14.1	29.5
Verkhoyansk	67°33′N 133°23′E	137	−46.8	−43.1	−30.2	−13.5	2.7	12.9	15.7	11.4	2.7	−14.3	−35.7	−44.5	−15.2	62.5
Kotel'niy	76°00′N 137°54′E	10	−29.5	−29.9	−27.0	−20.6	−9.1	−0.2	2.5	2.0	−1.5	−10.5	−21.9	−26.4	−14.3	32.4
Mys Schmidt	68°55′N 179°29′W	7	−25.4	−26.9	−25.1	−17.5	−7.2	1.5	4.0	2.9	0.0	−7.7	−16.3	−23.6	−11.8	30.9

Table 2.5 Locations, elevations and mean monthly and annual temperatures at selected Antarctic stations. Note that, to facilitate summer and winter comparisons with Arctic stations in Table 2.4, the columns are tabulated from July onward. Data from WMO (1971), Schwerdtfeger (1984) and Jacka, Christou and Cook (1984).

Station	Position	height (m)	Jul	Aug	Sep	Oct	Nov	Dec	Jan	Feb	Mar	Apr	May	Jun	Mean	Range
Plateau stations																
Amundsen–Scott	90°00′S	2835	−59.9	−59.7	−58.4	−50.7	−38.4	−27.7	−27.9	−40.2	−54.3	−57.3	−57.3	−58.2	−49.3	32.2
Vostok	78°27′S 106°52′E	3488	−67.0	−68.3	−66.3	−57.1	−43.4	−32.3	−32.3	−44.3	−58.0	−64.9	−65.9	−65.1	−55.4	36.0
Plateau	79°28′S 40°35′E	3625	−68.0	−71.4	−65.0	−59.5	−44.4	−32.3	−33.9	−44.4	−57.2	−65.8	−66.4	−69.0	−56.4	36.7
Byrd	80°01′S 120°00′W	1530	−35.6	−36.7	−36.6	−30.2	−21.4	−14.4	−14.7	−19.8	−27.7	−29.7	−33.0	−34.1	−27.9	22.3
Eights	75°15′S 77°06′E	420	−33.5	−37.0	−34.3	−28.3	−17.3	−11.3	−10.0	−18.0	−25.0	−31.1	−33.1	−33.5	−26.0	27.0
High-latitude coastal stations																
Little America	78°14′S 161°55′W	40	−36.3	−36.6	−36.7	−25.4	−15.6	−6.4	−6.6	−12.9	−21.8	−28.3	−30.6	−27.9	−23.8	30.1
Belgrano	78°00′S 38°48′W	50	−33.5	−32.9	−31.0	−21.7	−12.7	−6.1	−6.3	−13.2	−22.0	−26.2	−29.6	−32.0	−22.2	27.4
Halley	75°31′S 26°30′W	35	−28.9	−28.5	−26.2	−19.2	−11.6	−5.3	−4.8	−9.8	−16.6	−20.0	−24.4	−26.6	−18.5	24.1
McMurdo	77°50′S 166°30′E	24	−25.8	−26.9	−25.0	−19.5	−9.9	−3.8	−3.1	−8.8	−17.6	−21.1	−23.3	−23.5	−17.4	23.8
Hallet	72°18′S 170°18′E	5	−26.4	−26.6	−24.5	−18.3	−8.0	−1.7	−1.1	−3.2	−10.5	−17.8	−22.6	−23.0	−15.3	25.5
Low-latitude coastal stations																
Molodezhnaya	67°42′S 45°51′E	39	−18.0	−18.7	−17.8	−13.8	−6.6	−1.4	−0.5	−4.0	−8.3	−11.4	−14.2	−16.3	−10.9	18.2
Mawson	67°36′S 62°55′E	8	−17.8	−18.8	−17.7	−13.2	−5.4	−0.3	0.1	−4.4	−10.3	−14.5	−16.1	−16.8	−11.3	18.9
Mirny	66°33′S 93°01′E	42	−16.6	−17.4	−17.2	−13.5	−7.0	−2.4	−1.6	−5.3	−10.1	−13.9	−15.3	−15.6	−11.3	15.8
Oasis	66°18′S 100°34′E	28	−17.3	−16.4	−16.5	−11.2	−3.5	1.7	1.9	−2.3	−5.2	−7.5	−11.3	−20.5	−9.1	22.4
Dumont d'Urville	66°40′S 140°01′E	40	−16.2	−16.8	−16.1	−13.2	−7.0	−1.7	−0.7	−4.2	−8.7	−12.7	−14.7	−16.0	−10.7	16.1
Maritime Antarctic stations																
Orcadas	60°44′S 44°44′W	4	−10.5	−9.8	−6.4	−3.4	−2.1	−0.5	0.3	0.5	−0.6	−3.0	−6.7	−9.8	−4.3	11.0
Deception	62°59′S 60°43′W	8	−8.0	−7.7	−4.8	−2.4	−1.0	0.5	1.4	1.1	0.1	−2.1	−4.3	−6.3	−2.8	9.4
Hope Bay	63°24′S 57°00′W	13	−11.8	−10.6	−7.4	−4.1	−2.1	−0.2	0.2	−1.3	−3.7	−7.5	−9.3	−11.4	−5.8	12.0
Faraday	65°18′S 64°18′W	11	−10.7	−11.0	−8.3	−5.1	−2.6	−0.4	0.3	0.1	−1.0	−3.4	−5.8	−8.1	−4.7	11.3
Adelaide I.	67°48′S 68°54′W	26	−9.7	−11.4	−8.1	−5.3	−2.2	0.2	0.6	0.0	−1.5	−4.6	−6.5	−8.5	−4.8	12.0
Periantarctic and Subantarctic stations																
Grytviken	54°18′S 36°30 W	3	−1.5	−1.5	0.1	1.7	3.0	3.8	4.7	5.4	4.6	2.5	0.2	−1.5	1.8	6.9
Iles Kerguelen	49°12′S 70°12′E	12	1.9	2.0	2.2	3.3	4.7	6.2	7.2	7.7	7.1	5.7	3.7	2.3	4.5	5.8
Heard I.	53°06′S 72°31′E	4	−0.6	−0.8	−1.2	−0.2	0.5	2.1	3.3	3.5	3.0	2.4	1.3	−0.4	1.3	4.7
Macquarie I.	54°30′S 158°54′E	30	3.2	3.2	3.4	3.8	4.5	5.9	6.8	6.7	6.2	5.1	4.2	3.3	4.7	3.6

poorly, so very little penetrates beyond a surface veneer: thus surface temperature rises quickly by several degrees.

Radiation, conduction and convection losses from the warmed surface immediately increase, and a new dynamic temperature equilibrium appears, determined mainly by the rate of loss of heat from rock to air moving across it. In sheltered corners where the air is confined, a higher equilibrium temperature results for both rocks and air. Clouds passing across the sun reduce incident radiation and allow the surface temperature to fall; a slight increase in wind strength has a similar effect, and flakes of snow settling on the surface demand heat for melting and evaporation, causing a further fall. Each separate event produces a new equilibrium between rock surface and air.

Similar shafts of sunlight moving across snowfields produce a similar but lesser effect, for albedo is higher, more of the energy is reflected away, and temperature fluctuations are slighter. The highest temperatures occur in tiny air cells just below the snow surface; this is the microenvironment of cryoalgae (p. 107), which from time to time form large patches of red or green on melting snow surfaces. If the snow crystals warm enough to start melting, they change shape. This is a slow process because melting itself requires latent heat, and more heat is used to convert both ice and water to vapour. Snow and ice are important temperature buffers. In their presence ground temperatures remain below or close to freezing point, and not until they have melted in summer can the incoming energy be absorbed by rocks and soil. A November morning's sunshine in 78°S reduces the albedo of the snow surface from 80% to 60% by altering the reflective properties of the snow crystals, allowing greater absorption by the afternoon sun before albedo rises again in the evening (Thompson and Macdonald, 1962).

These changes are important to travellers over snow. Sledgers find surfaces harder when the sun is low; penguins crossing snow-covered sea ice to their nests save energy by travelling at night when the snow is firmest.

Spring thaw. The low albedo of overwintered snow allows thawing to start before air temperatures generally reach freezing point; hence the growth of icicles, made from water which has melted from snow but frozen again due to evaporation in contact with a subzero atmosphere. Hence too the melting and disappearance of winter snowfields in early spring, before monthly mean temperatures indicate that the thaw has started. During the same period of early spring lakes start to thaw at the edges, and green algae still encased in solid ice begin to metabolize. As pockets of frozen sand and shingle thaw at the surface, the interstitial water becomes available for terrestrial algae and soil organisms. Trickles of water run down the rocks and enter dry crevices, reviving organisms that have wintered there in safe desiccation.

Sunshine falling on thin layers of ice among rocks creates small temporary greenhouses, beneath which plants start to flourish in a warm humid atmosphere. Moss and lichen clumps soak up water and solar energy, gradually warming through to temperatures several degrees higher than ambient and starting their brief summer period of active metabolism. Mites, insects, nematodes and other microfauna among the stems and litter within the clumps are stir into activity, taking up their life-cycles where they left off in late summer. Though these changes can take place only when air temperatures are close to freezing point, their triggering depends almost entirely on direct effects of solar radiation, rather that on the movements of air masses and other factors that control the climate at large.

2.3 Weather and climate

2.3.1 *Variety in polar climates*

Climatic figures indicate that polar climates are typically cold, dry and windy. But polar regions are large and their climates range widely from severe to subtemperate. Greenland alone has climates ranging from intensely cold and dry in the northeast to mild and damp in the southwest. Continental Antarctica, over six times as large, has an even wider though less obvious range. The ice-cap, more than half of which is above 2000 m, is generally cold. But at Byrd Station, West Antarctica, mean summer temperatures in December and January are 13°C higher than at Amundsen-Scott (South Pole), and nearly 18°C higher than at Vostok, on the East Antarctica plateau. These differences, due mostly to altitude, are comparable with mean summer temperature differences between London, Morocco and the heart of the Algerian desert. Along the coast, between Scott Base in 78°S and Mawson in 68°S, mean temperatures for the coldest months differ by 16°C; this is greater than the difference in winter means between Athens and northern Norway.

Latitude and solar radiation are ultimate determinants of climate, but in both polar regions distribution of land, open water and sea ice are important factors too. Winter temperatures over sea ice in the north polar basin are higher than those over peripheral lands in much lower latitudes; even the highest points of the Greenland ice-cap, which are colder than the polar basin, are warmer than central Siberia in winter (Figure 2.7). In

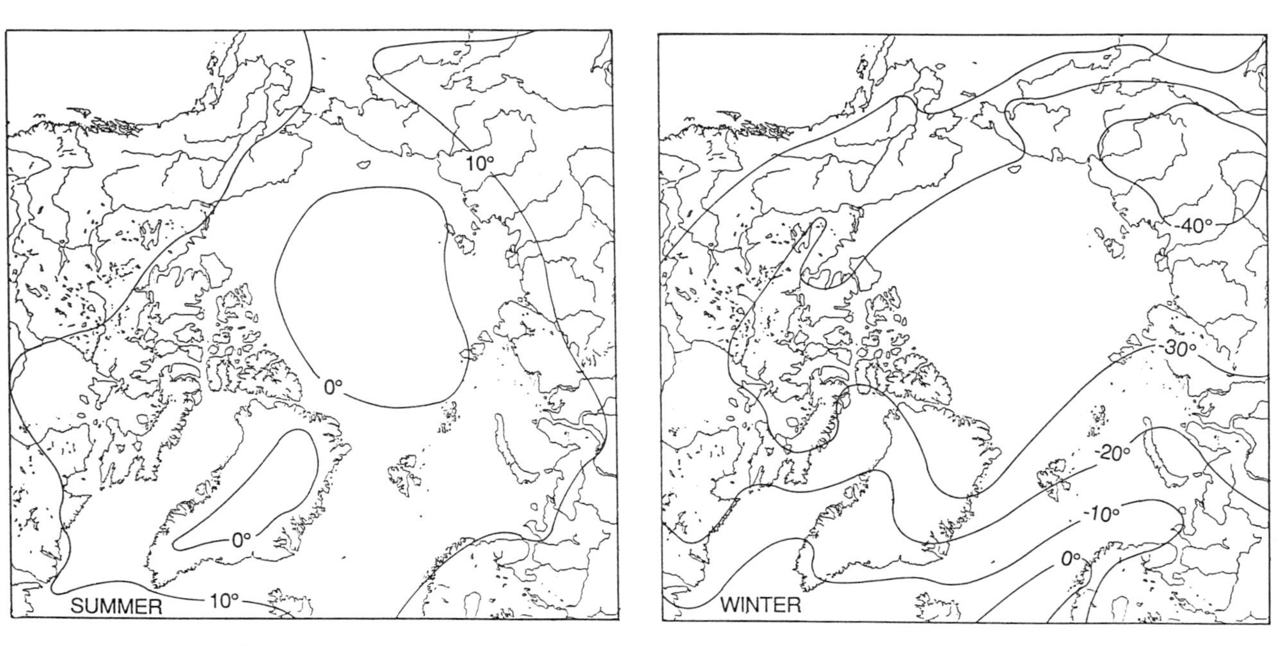

Figure. 2.7 Arctic summer and winter temperatures; isotherms of mean temperatures for the warmest and coldest months.

both polar regions living conditions may be better in high latitudes than low. There is more summer sunshine near the poles than in subpolar zones, where depressions bring cloud, rain and heavy snow. There are coastal stations that become completely buried under snow in two to three years, and high-latitude stations where precipitation is negligible and accumulation nil. Winds too are intensely local, with the strongest due more to local topography than to any other factor.

2.3.2 *Elements of climate*

Temperatures. Meteorologists are most concerned with air temperatures measured in screens and ground temperatures measured beneath the surface of soil or snow. Both are important for ecologists, especially if used in conjunction with microclimatic data; of the two, air temperatures are more readily available to show long-term trends and provide large-scale comparisons.

Weather stations in polar regions are few and far apart. Manned stations are particularly expensive to maintain in out-of-the way places, and increasing use is now made of automatic stations that transmit their data via satellites. Calculated or derived data are also used widely. For example monthly and annual air temperature isotherms (Figures 2.7, 2.8), are based on the few data available from stations, backed by interpolated values calculated from mean dry adiabatic lapse rates; for each 100 m rise in elevation temperatures fall 1°C up to 2000 m, 1.27°C from 2000 to 3000 m, and at higher rates in the thinner air above (Rusin, 1964). Pits dug in the accumulated snow and ice of glaciers can yield useful temperature data. At depths of 15–20 m, i.e. below the level to which diurnal and seasonal changes penetrate, snow temperatures approximate closely to the mean annual air temperature immediately above (Cameron and Goldthwait, 1961; Bull, 1964; Dalrymple, 1966), and are more readily measured.

Wind. Wind is air in motion; the motion is imparted by pressure differences due to differential heating of the atmosphere (systemic winds), or to topography (katabatic winds). Systemic winds include those prevailing in latitudinal bands—westerlies on the polar fringes and easterlies on their poleward flank—and the succession of shifting winds that accompany movements of depressions. Katabatic or downslope winds blow intermittently from high ground to low; the air usually warms and accelerates as it descends, and these provide some of the warmest and strongest winds in polar regions. When strong enough to lift and carry snow, winds create

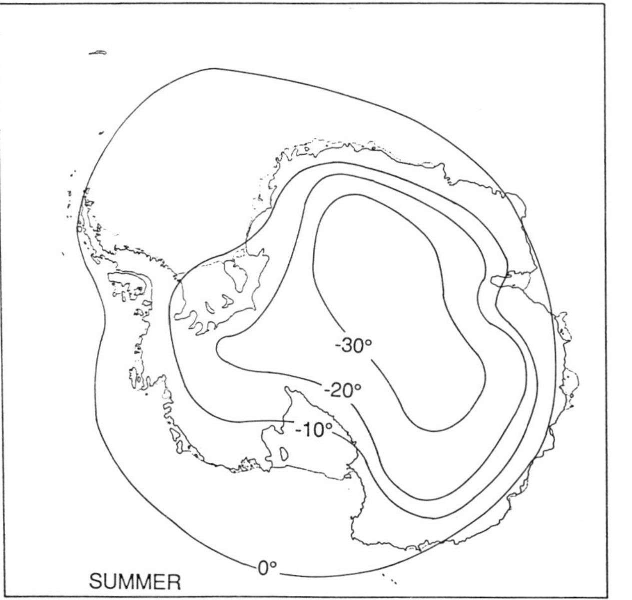

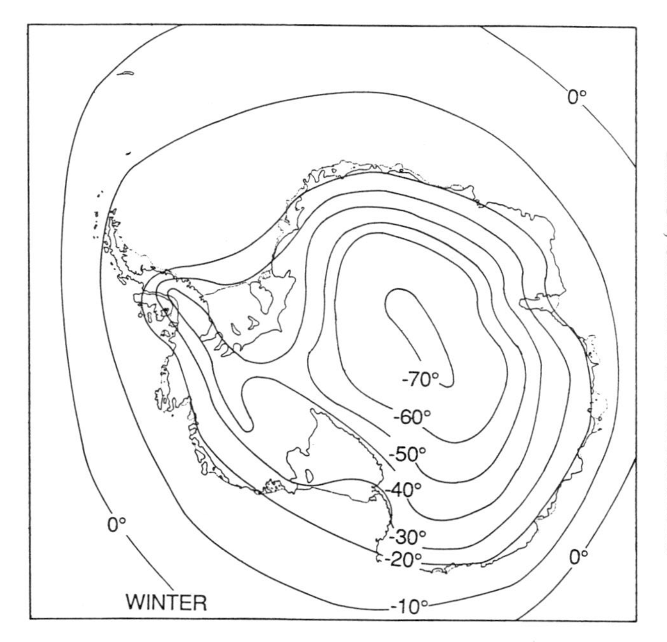

Figure 2.8 Antarctic summer and winter temperatures; isotherms of mean temperatures for the warmest and coldest months.

blizzard conditions, in which clouds of snow blow downwind often for days at a time. The snow is redistributed, sculptured into sastrugi and piled into hard-packed dunes which persist long after the general cover has melted in spring. Blizzards are a powerful erosive force, for at low temperatures the snow crystals become grit-hard, blasting rock faces and inhibiting plant settlement. They add substantially to the hazards faced by warm-blooded animals, including man, by obscuring visibility and increasing wind-chill effects; untimely summer blizzards may devastate breeding herds of caribou or colonies of seabirds (p. 98). Where vegetation is present, winds distribute seeds and other propagules, and thus help to spread organisms along their tracks. As systematic studies of 'aerial plankton' have shown, upper winds are important recruiting agents, transporting insects, mites, spiders and plant propagules constantly from temperate to subpolar and polar regions (Gressit, 1962; Yoshimoto *et al.*, 1962; Yoshimoto and Gressitt, 1963).

Humidity and precipitation. The chronically cold air of polar regions contains very little moisture; though relative humidity may be high, absolute humidity (the amount of water vapour in the atmosphere) is low, and the moisture available for living organisms is often meagre. Except where depressions bring warm, moist air into colder regions, precipitation is generally low. Many polar areas are dry deserts; others are semi-arid for long spells of the year when much of the ground water is frozen and neither rain nor snow are falling. Throughout polar regions lack of available water may be as limiting an ecological factor as extreme cold.

Precipitation figures are generally lacking from polar stations. Rainfall is easily measured in standard rain gauges, and snowfall, ice needles or hoar-frost caught in similar or larger gauges can be melted to provide a rainfall-equivalent (RE) value. However, snowfall is almost impossible to measure accurately if the wind is blowing, for winds above $7\,m\,s^{-1}$ whip up snow from the surface and pack the gauges, giving false readings. Better estimates of precipitation are gained by taking into account annual snow accumulation (Bull, 1971) and calculating monthly values by formula (Bryazgin, 1986).

2.4 Polar climates

Local climates and day-to-day weather are moderated by the movements of air masses that control air temperature, humidity and wind patterns in both polar regions. Air chilled over the poles becomes denser and sinks; air warmed over middle latitudes becomes less dense and rises. On a rotating earth, air drawn poleward from the temperate zone appears to spiral eastward and accelerate, flowing counter-clockwise about the Arctic and clockwise about the Antarctic. At ground level these airflows form the westerlies—the strong, persistent winds that dominate both subpolar zones. In the south they blow near-symmetrically about the continent between 40° and 55°S with only a slight seasonal shift. In the complex topography of the north they vary more in direction and strength, both locally and seasonally, diverted by persistent low- and high-pressure areas along their path.

2.4.1 *Air mass movements*

Depressions. Cyclones or depressions occur along the atmospheric polar front or boundary between cold polar and warm air masses from lower latitudes; the warm, moist air over-rides the cold and is forced upward into cool, overlying regions of the atmosphere. In the northern hemisphere depressions spiral counter-clockwise, in the southern hemisphere clockwise. They follow each other eastward around the world at rates of 600 to 1000 km per day, their progress monitored by weather satellites. Each depression brings a sequence of lowering cloud, rain or snow, sharp windshifts at the junction of air masses, followed by clearer, colder, showery weather.

Their eastward movements are diverted by other pressure systems. For example, persistent high-pressure areas keep them away from central Siberia and northern Canada in winter, diverting streams of depressions from the north Pacific to northeastern Asia and Alaska, and from the north Atlantic to Baffin Bay, Greenland and the Norwegian Sea. Their warm air brings mild winters and their heavy snowfall accumulates on the glaciers and ice caps. In the south the subantarctic islands owe their persistently dreary weather, with snow, rain and strong, shifting winds, to the depressions passing over them. South Georgia, Heard Island and the uplands of Iles Kerguelen are high enough to attract heavy snowfall and carry permanent glaciers.

High-pressure areas west of the Andes and southeast of Africa and Australia divert some depressions southward across the westerlies to the Antarctica coast on well-recognized routes (Taljaard, 1972). Those reaching Antarctica bring cloud and heavy snow, occasionally sleet and rain, especially to the Peninsula and neighbouring islands, Dronning Maud Land, Enderby and MacRobertson Lands, and the Wilkes Land and Terre

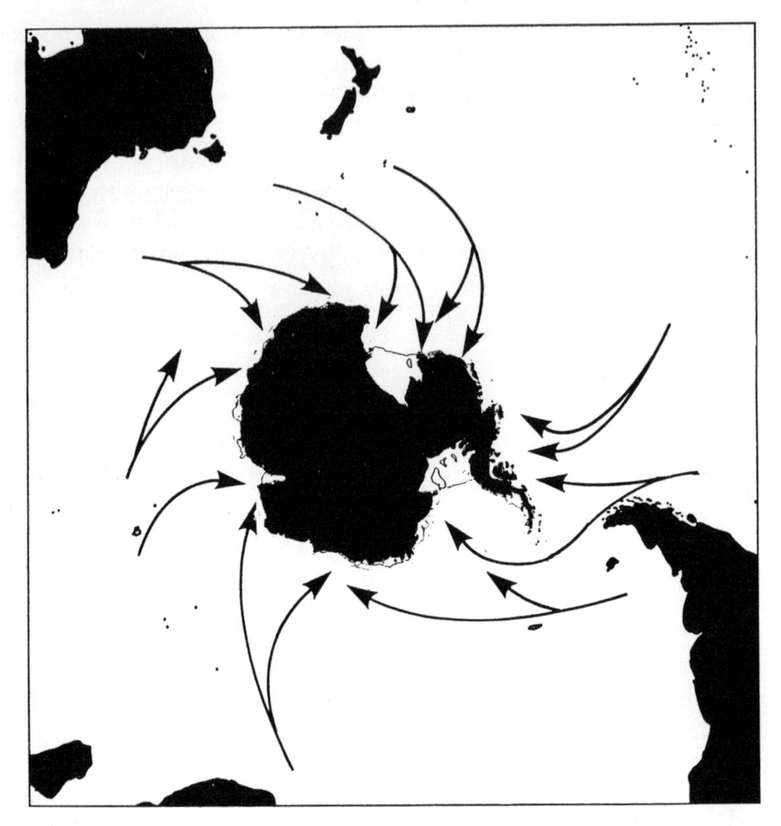

Figure 2.9 Paths of depressions across the westerly-wind zone to Antarctica; after Taljaard (1972).

Adélie coasts. Some cross the West Antarctica plateau; a few penetrate East Antarctica, occasionally as far as the South Pole (Figure 2.9).

Anticyclones. Anticyclones or high-pressure systems form on the warmer flank of the westerlies between 25° and 40° north or south, and also over the coldest areas of either hemisphere, especially in winter. Elliptical, often extended into the familiar 'ridges of high pressure', they are areas of clear, descending dry air, with an internal circulation that spins them clockwise in the north and counter-clockwise in the south. They form windows in the overcast, usually with a lifetime of two or three days but occasionally lasting much longer. Their clarity allows free flow of radiant energy to and from the earth; in summer they bring hot sunny weather, in winter clear days and intensely cold nights.

Transient anticyclones block and divert the passage of cyclones, often providing welcome breaks of good weather. Persistent winter anticyclones over Siberia and northwestern Canada make these the hemisphere's coldest and driest places. In Antarctica persistent anticyclones provide the sunny summers and clear, cold winters of the main continental area. Winds are light and clouds are high and dispersed; little snow falls, but there is often a slight, continuous fall of fine ice needles. Snow surfaces and the air adjacent to them cool more intensely than the air above, forming temperature inversions in the lowest layer of atmosphere. Inversions feature almost continuously over East Antarctica, except when disturbed by wind; those over West Antarctica are more often buffeted and displaced by cyclones moving in from the Weddell Sea.

Downslope winds. From the cold upland ice-caps of Antarctica and Greenland, notably during inversion conditions, the layer of cold air closest to the surface tends to be drawn downhill by gravity. The resulting winds follow the slopes closely, diverted slightly due to the earth's spin but controlled mainly by the contours. These are the typical light surface winds of both great ice-caps. On the steep coastal slopes they accelerate to produce some of the world's strongest and most persistent winds. Downslope winds blow almost everywhere along the coasts, especially where the inland topography channels them, forming a layer only 100–200 m deep and seldom extending more than a few kilometres beyond the shoreline (Figure 2.10).

	MS1	MS2	MS3	MS4
Distance from coast (km)	25	10	0.7	13.5
Height above sea level (m)	600	400	0	0
% downslope (SE, SSE, S) winds	74	88	74	71
Mean wind velocity (m sec^{-1})	9.8	10.8	11.3	4.4

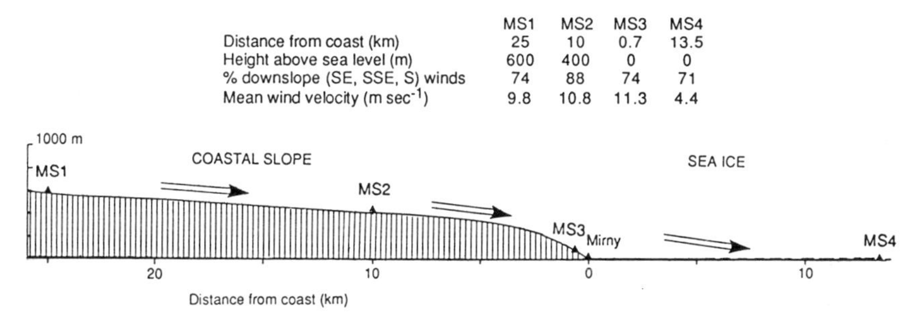

Figure 2.10 Katabatic or downslope wind, measured at three mobile stations (MS1–3) on the Antarctic icecap above Mirny, and one station out on the sea ice (MS4). The wind reaches its greatest velocity at the coast, then loses momentum rapidly over the sea ice. Data from Tauber (1960).

The strongest and most constant winds blow in the Terre Adélie sector of East Antarctica, which is probably the windiest place on earth. At Cape Denison in 1913 Sir Douglas Mawson's expedition (Madigan, 1929) recorded a mean annual wind speed of 19.4 m s^{-1} (43.40 m.p.h.), almost entirely south or southeasterly. The mean for the quietest month (February) was 11.7 m s^{-1} (26.17 m.p.h.), for the stormiest month (July) 24.8 m s^{-1} (55.48 m.p.h.), and for the stormiest single day (16 August 1912) an astonishing 36.0 m s^{-1} (80.53 m.p.h.). Similar winds have more recently been reported by French expeditions at Port Martin some 60 km westward along the coast (Loewe, 1972). The causes and mechanics of these exceptional winds are discussed by Schwerdtfeger (1984).

2.4.2 *Arctic climates*

Climatically the Arctic can be divided into a central maritime basin, and the lands and seas peripheral to it. The central area has no permanent meteorological observatories, but has been studied at a succession of scientific stations drifting on ice islands. The peripheral area has sparse but long-term coverage from a ring of observatories; mean temperatures from selected stations appear in Table 2.3.

Ocean basin. The central core of pack ice is climatically the most stable area. In the cold, sunless winter a strong anticyclone forms centrally. Clear skies and light centrifugal winds prevail over the shifting ice; the sea is almost completely covered and well insulated by the floes, many of which are over 3 m thick. Temperatures average −30 °C or lower during December, January and February, and −26 °C to −28 °C along the coasts. Atmospheric inversions often occur; in recent years a haze of industrial pollution has become evident in the lower atmosphere, drifting in from distant cities and smelters on the mainlands (Stonehouse, 1986; for an overview see Barrie, 1986).

When the sun returns the anticyclone weakens and is invaded by peripheral depressions, which bring cloud, moist air, fog, snow and rain. Temperatures over the pack ice rise to freezing point or above in all but a small central core of the ocean basin. Winds strengthen and vary as the depressions pass. The edges of the pack ice melt during the warmest months of June and July, and widespread open water appears along the coasts of Alaska and Eurasia. Air temperatures on the Siberian islands (Kotel' niy and Ostrova Novosibirsk, Table 2.4 and Figure 2.11) and mainland coasts (Dikson, Mys Chelyuskin, Mys Schmidt) rise a degree or two above freezing

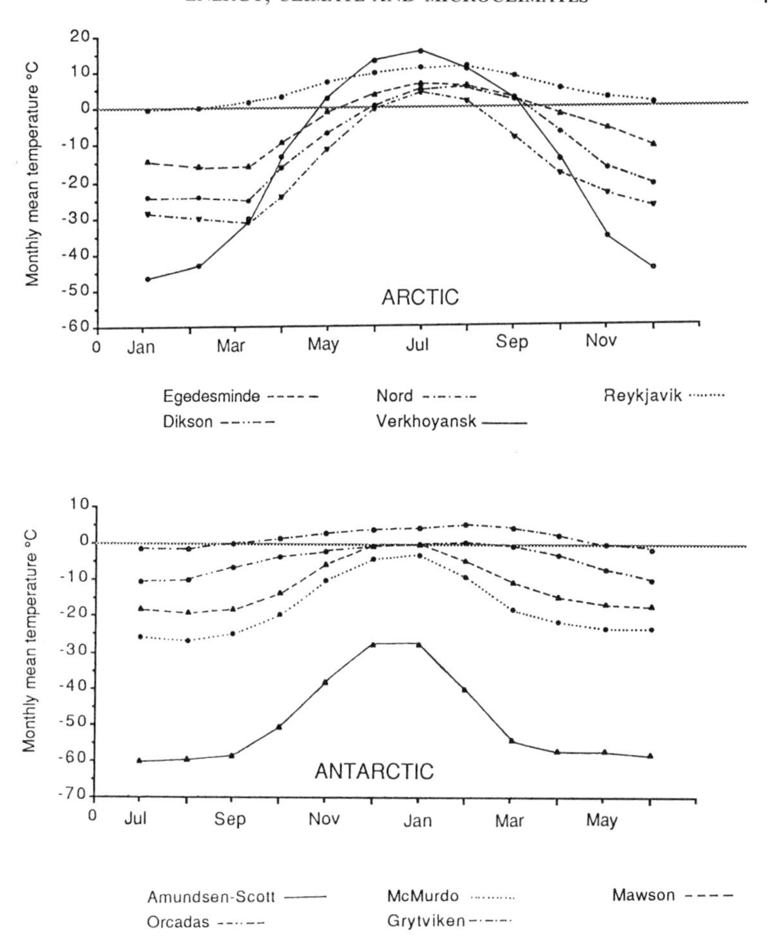

Figure 2.11 Monthly temperatures at selected Arctic and Antarctic stations; data from Tables 2.4 and 2.5.

point, snow disappears and the ground thaws superficially, allowing the meagre flora and fauna its spell of summer activity.

North America. The persistent winter anticyclone spreads cold air over much of central and northern North America from November to May; the coldest months are January and February, and the coldest sector is the central Canadian north, here represented by Inuvik on the mainland and Resolute, Cornwallis Island. Winters are similar to those in the central polar basin, with clear or thinly-veiled skies (often enlivened by brilliant displays of the aurora borealis), light winds and slight precipitation. Persistent temperature inversions stabilize the atmosphere, causing clouds of smog to form over many settlements (Benson, 1986).

Summers are short; the thaw starts as the sun gains strength in March on April, but mean monthly temperatures above freezing point are rare before May. As the winter anticyclone weakens, depressions swing in from the south, bringing cloud, snow squalls and quick shifts of temperature. Precipitation is highest in the northwest and northeast; warm moist oceanic air contributes snow to the glaciers of Alaska and the ice-caps of the eastern Canadian archipelago. June, July and August are the warmest months; as the snow clears and the ground warms, air temperatures rise well above freezing point and remain high during the 24 hours of daylight. There is more cloud, and snow, sleet and rain keep the atmosphere moist. These conditions support a flourishing tundra vegetation over much of the ice-free area.

Greenland, Iceland. Greenland's massive ice-sheet, rising to over 3000 m, is a permanent centre of cold. Maintained by snow from year-round moist southwesterly airstreams, its high plateau surface is one of the Arctic's two coldest areas in winter (see below), with mean temperatures of $-40\,^{\circ}C$ to $-45\,^{\circ}C$. Summer temperatures reach $-12\,^{\circ}C$ along the high central ridge. The snow that maintains the ice-sheet falls mainly from depressions, reaching 100 cm RE annually in the south but only about 20 cm RE annually in the north. Almost ice-free, the north end of the island is dry and barren.

The north shore (Nord, Table 2.3) is blocked by semi-permanent sea ice. Mean monthly temperatures fall below $-30\,^{\circ}C$ in winter and are above freezing point only in July and August. The east coast (Angmagssalik) and northern half of the west coast (Egedesminde) are dominated by cold southward-flowing currents which keep sea ice abundant even in summer. The southwestern corner (Prins Kristian Sund), warmed by a branch of the North Atlantic Drift, is relatively free of sea ice and much milder throughout the year. Here are found rich tundra, grazing grounds for sheep and thickets of near-forest vegetation.

Most of Iceland is subarctic, with mean monthly temperatures above freezing point throughout the year. Sandwiched between the warm North Atlantic Drift and the cold East Greenland current, its southern shores are mild and damp (Reykjavik), while its northern shores (Akureyri) are colder and often invested by drifting pack ice. Depressions bring rapidly changing weather to the island, and year-round snow to the south and east, filling the mountain valleys with permanent ice. Wet tundra and grasslands flourish, with stand of birch and spruce forest in the south.

Svalbard and northern Norway. The warm current that tempers the climates of southern Greenland and Iceland sweeps on toward Svalbard, bringing anomalously warm conditions to Jan Mayen, Bjørnoya, the Svalbard archipelago (Isfjord) and the northwestern tip of Europe (Vardø, Bodø). Though all the islands see both pack ice and fast ice in winter, only the north coast of the archipelago is ice-bound in summer and correspondingly cold. On the Norwegian mainland both Vardø and Bodø are year-round ports, despite their positions well north of the polar circle. Depressions that sweep in from the southwest bring snow in winter and rain in summer; only the larger islands of the archipelago have substantial ice-caps. Tundra covers the islands, forest-tundra the mainland localities.

USSR. Nearly all of the long arctic coast of Soviet Europe and Asia lies north of the polar circle, reaching a northernmost point beyond 77°N at Mys Chelyuskin. Most of it comes under the influence of the intense winter anticyclone centred about Verkhoyansk (Table 2.3) in Siberia, from which cold, dry air spreads in all directions. Verkhoyansk itself, well away from ameliorating influences of the sea, holds the northern hemisphere record for extreme cold (−67.8°C in January). It probably comes close to the world record for extreme year-round temperature range, for the central Siberian summer is short but hot; July temperatures over 36°C have been recorded, yielding an annual range of more than 100°C. Precipitation is low, falling mostly as rain in summer and totalling only 15 cm for the year.

The western edge of the Arctic (Murmansk, Archangel'sk) is warmed throughout the year by the North Atlantic Drift, which extends as far eastward as the eastern shore of Novaya Zemlaya and keeps the Barents Sea open in summer. Depressions bring winter snow and more abundant summer rain, mostly between June and September, averaging 35 to 55 cm per year. Dikson, a port on the Yenisey estuary, represents more fully the continental influences that dominate the remainder of this coast, though its climate is softened by the outflow of warm water from its great river. Precipitation is similar but less, totalling 26 to 30 cm annually. Turukhansk, almost 1000 km south of Dikson along the same river, has a warmer and more continental climate typical of the inland arctic fringe, with abundant rain in summer. Further east (Mys Chelyuskin, Kotel'niy, Mys Schmidt) conditions again become harsher, with short chill summers and intensely cold winters. Mys Schmidt and Mys Chelyuskin respectively have 29 and 24 cm RE annually; Kotel'niy, on Ostrova Novosibirskiy, is a true arctic desert with only 13 cm of precipitation per year.

2.4.3 *Antarctic climates*

The plateau. The high plateau of East Antarctica and lower plateau of West Antarctica are the coldest regions of the continent on record; only the high peaks of the Ellsworth Mountains are higher and likely to be colder, especially in winter. In Table 2.3 Vostok and Plateau stations represent the highest inhabited points of East Antarctica; the plateau itself rises slightly higher, to over 4000 m. An intermediate level is represented by the south polar station Amundsen-Scott, and Byrd represents the lower West Antarctica plateau. Eights station stands lower down on the plateau edge, at its junction with Antarctic Peninsula.

Vostok, operational since 1958, recorded the world's lowest temperature of – 88.3°C on 24 August 1960; temperatures very close to this occur every year about the same time. Plateau station, slightly higher but operational only for the three years 1966-68, recorded lower mean annual and monthly temperatures, and would probably have provided a lower record temperature had it remained open longer. Amundsen-Scott, 620 m lower than Vostok and operational a year longer, is on average 3-5°C warmer in summer and almost 10°C warmer in winter.

Temperatures at all these stations fall rapidly between February and April, then level off for the four or five winter months following. There is often a slight fall in late August or early September just before the sun reappears. This curious flattening of the annual temperature curve (Figure 2.11) typifies the kernlos or coreless winter, caused by persistent inversions that reduce heat losses from the snow surface (Schwerdtfeger, 1984). The annual range between warmest and coldest months exceeds 30° C at the higher stations and 20° C at the lower ones.

However well dressed, man finds it hard to work in temperatures below – 50°C. Living conditions outdoors on the high plateau are difficult, but some polar workers find the constant summer sunshine, light winds (mean 4–5 m s^{-1}) and low precipitation (3–5 cm RE annually at Vostok, 6–7 cm at Amundsen-Scott) more tolerable than the warmer, cloudier and windier conditions on the coast (p. 184). Summer temperatures on the plateau compare with winter temperatures in the inhabited Canadian north; the mean for the warmest month at the South Pole (– 27.4°C) is the same as that for the coldest month at Churchill, Manitoba. The cold that bedevilled

Scott and his companions on their march from the Pole in 1912 (Scott, 1914) would have been quite familiar to the inhabitants of Winnipeg, Regina, Saskatoon, and many Soviet arctic cities.

The West Antarctica plateau (Byrd, Eights) is warmer with more depressions, more cloud, heavier snowfall and stronger winds. At Byrd snow level rises about 30 cm annually (enough to bury a surface station in five or six years) with an annual RE of 14–16 cm. Eights station, lowest and northernmost of the plateau stations and occupied for only three years (1963–65), occasionally recorded summer temperatures above freezing point.

Ecologically the plateau areas are deserts; neither plants nor animals are permanent inhabitants, though they are visited occasionally by skuas (*Catharacta maccormicki*) and other birds. Nunataks and mountains fringing the plateau support a meagre flora and fauna almost to their southern limit (p. 88). For much of the year these share the coldest conditions of the ice plateaux; only for brief periods in summer can favourable microclimates develop around them.

Continental coasts. The southernmost sectors of antarctic coast lie in 78°S, the northernmost in 65°S; not surprisingly, there is a wide range of coastal climates. Latitude is but one contributing factor; others include the frequency of depressions, the inland topography (influencing strength and direction of prevailing winds), and the presence or absence of sea ice offshore in summer. The coldest shores are in latitudes south of the 70th parallel where depressions rarely penetrate, snow-cover persists the year round, and thick bay ice or pack ice last through the summer, insulating the coast from warming effects of the sea (in Table 2.3, Little America, Ellsworth and Halley). Slightly warmer in summer are McMurdo and Hallet, high-latitude stations on raised beaches bordering the Ross Sea, where sea ice persists but snowfall is light and rocks are exposed for three or four months each year.

These high-latitude coastal stations have continental climates similar to those on the lower plateau, with winter mean temperatures below −24°C, annual means below −15°C, and a range of over 20°C between means for the warmest and coldest months. Winds average about 6 m s^{-1}, the strongest usually downslope. Cloud is thin, and summers are almost as sunny as those on the plateau.

Coasts in lower latitudes are warmer and cloudier, with frequent depressions, stronger winds (means 10–12 m s^{-1}), and mean annual temperatures close to −10°C. Bay ice forms in autumn and persists through winter, reaching 1–2 m thick and blanketing the sea; winter means of −20° to −15°C are common. The sea ice disperses in November or December, allowing air temperatures to hover around freezing point for three or four summer months. Often the ground becomes snow-free, allowing further warming for as long as the open water persists. The annual range between means for the warmest and coldest months is usually less than 20°C, reflecting the strong maritime influence. Among the five stations illustrating this group in Table 2.3 snow-free Oasis is anomalous; located a few kilometres from the sea, with the highest summer and lowest winter mean temperatures (see above), it has a continental temperature range.

Precipitation ranges from about 20 cm annual RE on drier southern coasts to 40 cm or more—representing 2–3 m of fallen snow—where precipitation is heaviest. Coastal stations built on ice shelves generally find themselves in high accumulation areas and liable to disappear under the snow after a few years. Those on beaches are in sites where annual precipitation is low or matched by summer melting, and have a longer span of usefulness. Rain falls often in summer at the more northerly stations.

Seabirds and seals are distributed patchily along all these coasts; even the southernmost icecliffs provide shelter for emperor penguins *Aptenodytes forsteri* and Weddell seals *Leptonychotes weddellii*, which feed through cracks in the sea ice. Indeed some birds fly beyond the coasts to breed on mountain tops many kilometres inland (p. 154). Terrestrial ecology is poor along the high-latitude coasts, where plant communities seldom rise above a polar desert level. In lower latitudes, with longer summers and more melt-water available, plant life and its contained microfauna become very much richer, though neither brown soils nor flowering plants have so far been found even in the warmest spots along the continental coast. Though snow-free and mild in summer, 'oasis' areas lack water and are usually dry deserts (p. 90), with virtually no vegetation.

Maritime and peripheral Antarctica. The west coast of Antarctic Peninsula with its off-lying islands, and the South Orkney, South Shetland and South Sandwich archipelagoes, are climatically and ecologically better-favoured than the continental coast. The peripheral islands close to the Antarctic Convergence, lying outside the zone of pack ice, are most open to year-round oceanic influences, and ecologically more favoured still. Climatologically these are also the best-recorded areas. Observations have been maintained continuously at Orcadas (South Orkney Islands) since 1903, at Grytviken (South Georgia) since 1905, and at Deception (South Shetland Islands) and several stations along the Peninsula since World War II.

Reaching well beyond the Antarctic Circle, the peninsula stands in the track of depressions which, winter and summer alike, provide plenty of cloud and snow. Investing sea ice ensures cold winters of almost continental severity, but its dispersal provides summers of a mildness unmatched on continental Antarctica. Maritime conditions extend poleward to about 64°S on the colder Weddell Sea flank and to 69°S in the west. Because of Weddell Sea influences, Hope Bay at the northeastern tip of the Peninsula is marginally colder than Adelaide Island far to the southwest, and the South Orkney Islands are colder, with shorter summers, than the South Shetland Islands.

Along the west coast there is a gradient of increasing annual precipitation from 30 cm RE in the south to over 100 cm RE in the north. Each year rain is recorded some 10 days in the south, 54 days at the Argentine Islands (Faraday) and over 100 days in the South Orkneys (Pepper, 1954). The whole peninsula is glaciated, with the northwestern islands capped by domes of ice and soft snow. However, melting during the relatively warm summers ensures that there is open ground and plentiful water available for plant life. Thick, peaty accumulations of moss are common on damp, sunny slopes, soils reach levels of maturity unmatched on the continental coasts, and plant communities (including Antarctica's two species of flowering plants) are richer and more varied than at any continental locality.

The peripheral islands have no investment of pack ice, and at sea level take their mean temperatures from the Southern Ocean itself. South Georgia is alpine and heavily glaciated, Heard Island has a single glaciated peak; the main island of Iles Kerguelen has upland glaciers amongst its highest mountains, and Macquarie Island is unglaciated. Cloudy, windy and wet, with plentiful snow in winter, they support lush grass, herb and moss meadows on the coastal lowlands and tundra-like fellfields close to the snowline above.

Microenvironments and microclimates. Within climatic regions, local climatic conditions vary widely over very short distances; warm sheltered valleys alternate with harsh exposed hilltops, and sun-traps are backed by cold, sunless corners. During studies of Truelove Lowland, a coastal area of Devon Island in the Canadian Arctic, as part of the five-year International Biological Programme of 1967–72 (Bliss, 1977), ecologists began with five microclimatic stations but soon found the need for seven more, to cover their relatively small but topographically varied study area of 43 km^2 (Courtin and Labine, 1977). The lowland, a well-vegetated patch within an area of polar desert, owed its lush vegetation to a marginally warmer local climate; this allowed a growing season some three weeks longer than on the neighbouring uplands, more persistent moisture throughout summer and slightly warmer soils. Sites most favourable to plant growth included west-facing limestone cliffs that caught the afternoon sun, and a low hill warmed by air coming from the mouth of a sun-warmed canyon; both were blown free of snow in spring, allowing the longest possible growth season.

Many studies of microclimates in soils and moss clumps have been made on mainland Antarctica (for example Prior, 1962; Janetschek, 1967; Matsuda, 1968), in maritime Antarctica (Collins, 1977; Tilbrook, 1977; Walton 1982a and b, Friedman *et al.*, 1987) and on South Georgia (Walton, 1984). These agree that in general terms the climate close to the ground is more benevolent to living organisms than that measured in screens. In sunshine, even hazy sun seen through cloud or thin layers of ice or snow, mean surface temperatures are usually several degrees higher than air temperatures. In vegetation and among rocks that give shelter from wind, both temperature and humidity are likely to be higher. Even with ambient air temperatures well below freezing point, soil temperatures may remain a few degrees above freezing for days or weeks on end in summer, and conditions for growth among the stems, thalli and leaves of well-established plants may approach those found in temperate regions.

One most comprehensive study of antarctic microenvironments is the three-year programme of monitoring by satellite undertaken by Friedman and others (1987) at 1650 m above sea level in the Asgard Range of South Victoria Land. Investigating microclimates on dry sandstone Linnaeus Terrace (77°36′S), these observers found that conditions in which significant photosynthesis could occur, represented by the number of hours above freezing point at different microenvironments around their site, extended from mid-November to early March and ranged from 50 to 550 hours annually, depending on orientation.

2.5 Organisms at low temperatures

Biological processes are chemically based and subject to the rules governing chemical reactions. A most important rule is that covering temperature; rates of all chemical reactions, whether biological or not, are temperature-dependent, occurring more slowly at low temperatures than at high. Within biological systems the rates of most reactions increase by a factor of 2 to 3 for each 10 °C rise in temperature. This is usually expressed as a Q_{10} relationship; the velocity constant K_1 of a reaction at temperature t_1, and the velocity constant K_2 of the same reaction at temperature t_2, are related by the expression:

$$Q_{10} = \left(\frac{K_1}{K_2}\right)^{10/(t_1 - t_2)}$$

This relationship for living systems holds good within narrow limits, but is severely upset by temperature extremes, mainly because biological reactions are subject to another level of temperature-sensitivity. Many intracellular reactions are enzyme-catalysed, and the enzymes that moderate the processes often work optimally only within a narrow range of temperatures. Nearly all, for example, are slowed almost to zero at temperatures near to freezing point, and completely inactivated at temperatures around an upper limit of 41°C; some work only within much narrower ranges of a few degrees Celsius. Thus Q_{10} values vary considerably from one 10°C range to another; for a useful discussion of temperature in biological systems see Hardy (1972).

In practice it generally pays organisms, large or small, to work within a narrow range of body temperatures. Both plants and animals have the means of optimizing and stabilizing their working temperatures.

2.5.1 *Poikilotherms*

Poikilotherms (cold-blooded organisms) include most animals and all plants. With few exceptions the working temperatures within their bodies at any moment are dependent on the temperature of their environment, and throughout the world they work at a very wide range of temperatures from well over 40°C in hot deserts to well below freezing point in polar regions. Solar radiation is an important factor in their lives; it may be the first factor that raises their temperatures from low ambient to effective working range. Hence it does not generally pay poikilotherms to be insulated, though a few (for example bumblebees) have limited capacity to retain self-generated heat.

Aquatic poikilotherms run almost entirely at the temperature of their environment (Chapter 2). A few species of fast-moving predatory fish of temperate and tropical seas have anatomical adaptations that allow them to raise and maintain the working temperatures of their swimming muscles several degrees above ambient, but most poikilotherms cannot. Polar fish and invertebrates work at the near-freezing temperatures of their environment, and appear to be at no disadvantage in doing so; indeed by living at lower temperatures they use less energy and may live longer. Many, though not all, work only within very narrow temperature ranges, and are incapacitated if their ambient temperature rises more than a fraction of a degree above normal.

Most terrestrial poikilotherms are able to attain working temperatures that differ favourably from those of the environment—animals through behaviour, and plants through kineses or small body movements. In all regions of the world this is generally a matter of raising the body temperature a few degrees above environmental temperatures, especially in the mornings when both plants and animals need to recoup their overnight radiation losses. In polar regions active life may be restricted to slopes where summer sunshine allows sufficient warming for a few days or weeks, to be replaced by thermal torpidity and inactivity during a much longer winter. Overheating may occur in summer sunshine, especially in dry areas; shortage of free water restricts many plants to xerophilous form and water-saving habit (Chapter 3).

2.5.2 *Homeotherms*

Homeotherms (warm-blooded organisms) maintain high and constant body temperatures. Though a few species of tropical reptiles contrive to live almost homeothermous lives, only birds and placental mammals are true homeotherms, able to maintain steady core temperatures of 39–41 °C against steep environmental temperature gradients (Figure 2.12). The advantage of homeothermy is presumably the internal stability that had allowed birds and mammals to develop, for example, their sophisticated central nervous systems. The disadvantage is cost; it takes a deal more energy to run a homeotherm in any but the warmest climates. They require thermal insulation, plentiful food, and often an internal energy store to allow for erratic food supplies.

Despite the great differences between body temperature and environmental temperatures, homeotherms are well represented in polar climates. Especially plentiful are aquatic or semi-aquatic birds and mammals, for example polar bears, penguins and petrels. Well insulated with thick fur, feathers or subcutaneous fat for life in cold water, they seem pre-adapted for life in polar terrestrial environments as well (Chapter 6). However, the Arctic especially has many other species less obviously pre-adapted, including very small rodents and finches which appear to have spread fairly recently to the polar environment and found niches that they can occupy successfully. Though hibernation is often considered an energy-saving strategem for living in cold climates, in fact no polar homeotherms hibernate and few even become torpid. Female polar bears at denning (breeding) are unusual in entering a limited winter sleep, allowing their

Figure 2.12 Polar homeotherms; Weddell seal with pup. At birth in the Antarctic spring the ambient temperature of a pup may fall from 39°C in utero to −30°C or lower outside. Photo: Guy Mannering.

Figure 2.13 Microclimate: measuring wind strength close to the ground on a colony of Adelie penguins. Gales are much reduced by friction at ground level. Photo: John Darby.

body temperature to fall only a few degrees for a few days at a time (Larsen, 1978). Many homeotherms take advantage of mobility to migrate into polar regions for a period of summer plenty, but winter in warmer areas.

2.6 Summary and conclusions

Solar radiation is the mainspring of climatic processes all over the world. Polar regions are cold because incoming radiation is unevenly distributed throughout the year, and much is lost from the highly reflective sufaces of ground and ocean in high latitudes. In polar environments radiation controls weather processes, but additionally, in the absence of tall vegetation and other buffering factors, it plays a more direct role in determining the living conditions of plants and animals at ground level. Analysing the balance between incoming short-wave radiation and outgoing long-wave radiation at three arctic and three antarctic stations highlights some of the factors that are important for living organisms in polar microenvironments. On a broader scale, living conditions are dictated primarily by temperature, wind and precipitation. While both polar regions are cold, both are large and geographically varied, enough to show a wide variety of climates, many of them tolerable for living organisms. The climates of both polar regions are reviewed, with special reference to their suitability for supporting life. The effects of temperature on chemical reactions are considered, and the strategies of homeothermy and poikilothermy are briefly reviewed.

CHAPTER THREE
TERRESTRIAL ENVIRONMENTS

3.1 Introduction: polar land environments

Polar communities of plants and animals testify to their own hardiness and adaptability, but also to positive advantages of polar living that ecologists from warmer climates tend to forget. Inherent disadvantages of cold, aridity, immature soils, snow, permafrost and poor drainage must be amply compensated, for example by relative freedom from interspecific competition, parasites and predators; if there were no such compensations, polar lands would lie barren. Seldom is polar vegetation so lush as subpolar or temperate vegetation, nor its productivity per hectare so high; arctic tundra, at first glance green and meadow-like, is often the thinnest of carpets with bare rocks and shingle showing through. But both polar regions support vegetation and land fauna that are well adapted to their environments (Chapter 6) and varied enough to keep generations of ecologists interested in them. The apparent simplicity of polar land ecosystems gives them a special value for modelling and quantitative studies.

Origins and history of polar biota. Tropical and temperate environments have existed continuously on earth for hundreds of millions of years; polar conditions are relatively new. From the end of the Permo-Carboniferous ice age 250 million years ago to about 20 million years ago, polar regions have been without permanent ice. The species of plants and animals that inhabit them today have adapted, where adaptation was needed, only during the current ice age. Toward the end of the Pliocene epoch 2–3 million years ago the earth's coldest environments were alpine. These are the most likely sources of today's polar terrestrial biotas. Only a few species inhabit both polar regions; each region has recruited almost entirely from its own hemisphere.

The environment in which polar adaptation occurred was a turbulent one, conducive to rapid evolution of a small number of hardy species. Both

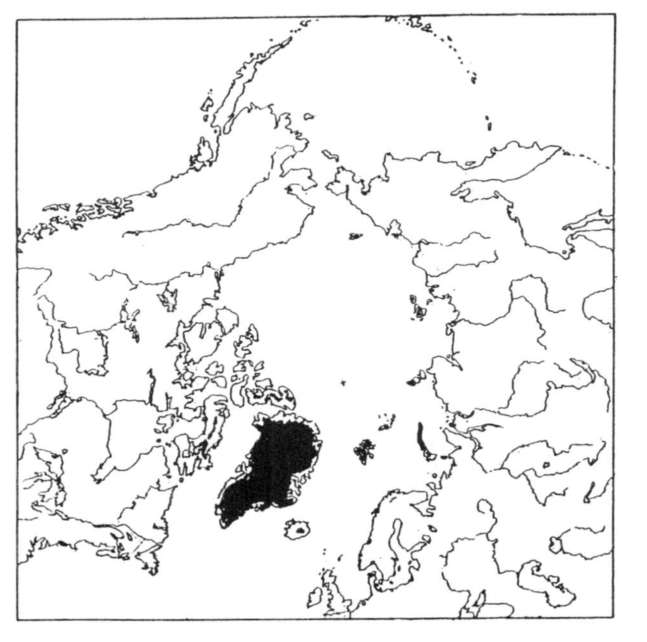

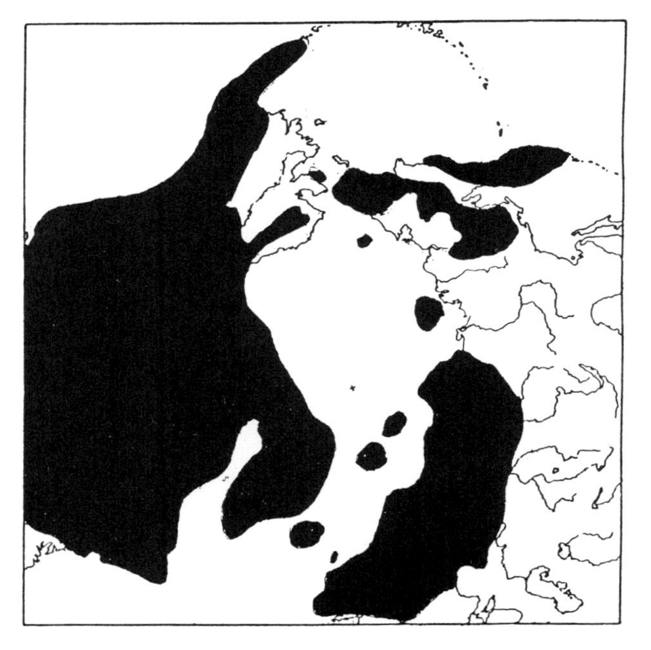

Figure 3.1 Greatest extent of land-based ice in the northern hemisphere at present (left), and at its widest extent during the present ice age (right).

polar regions show evidence of more extensive glaciation in the recent past. The shape, extent and histories of successive ice-sheets have been deduced from moraines, ice-scratched rocks, eskers, drumlins, altered drainage patterns and other periglacial features. Sedimentary cores from the sea-bed, dateable by oxygen isotope measurements of their microfauna, show the spread and retreat of cold oceanic water, and hence the extent of the ice-caps (Hays *et al.*, 1976; Emiliani, 1978). Pollen successions from lake-beds and cave-floor bone deposits, often dateable by radiocarbon methods, mark the ebb and flow of periglacial biotas; for a review of the European flora and fauna during this period see Kurtén (1968).

The current glacial period. The concept of four major glacial events, based on classic studies in the European Alps (Penck and Brückner, 1909) and supported by observations in Scandinavia, North America and Siberia, has in the last decade been shown to be an oversimplification. Evidence from sea-bed sediment cores suggests that, in the 1.8 million years since the start of the Pleistocene, the temperate and polar regions have passed through a dozen or more warm–cold cycles caused primarily by changes in levels of incident radiation. Each cycle, lasting about 100 000 years, was marked by extension and build-up of ice, followed by a relatively rapid deglaciation. For a review of these findings and the evidence for them see Imbrie and Imbrie (1979).

In the north during the most recent cycle the land ice margin retreated and advanced, for example, over the northern plains of Canada and the USA from northern Ontario and Quebec to Indiana and Ohio. There were major advances some 60 000 and 40 000 years ago, but the greatest expanse of ice during this cycle occurred about 20 000 years ago (Goldthwait *et al.*, 1965), when much of the present arctic tundra and wide areas of subarctic and temperate fringe lands lay under ice (Figure 3.1).

Thus, particularly in the north, this was a period of shifting climates and habitats, in which forest, tundra and ice-sheets replaced each other repeatedly and rapidly. The polar habitats we see today are likely to be just as transitory as those that preceded them.

3.2 Deglaciation: soil formation

3.2.1 *Shedding the ice*

Northern deglaciation. In the north deglaciation has accompanied a slow, unsteady warming, in which the overall loss of ice has been punctuated by spells of recovery. With each warm spell tundra and forest edges have moved north, only to be forced south again during cooler spells. Some 10 000 to 12 000 years ago the northlands were much colder than they are today, with mean temperatures 3–4°C lower. Nine thousand years ago they were about as warm as today, though there was more ice present. Six to seven thousand years ago a warm spell, the 'climatic optimum', brought southern European forests to Britain and dispersed ice from the Arctic Ocean. Five thousand years ago, and again three thousand years ago, the chill returned, the pack ice re-formed, and the northlands were colder than they are today (Imbrie and Imbrie, 1979).

Since then they have been both warmer and colder, with oscillations of the order of 1–2°C about annual mean temperatures. Though small on a world scale, these shifts represent milder or harsher winters, more or less precipitation, longer- or shorter-lasting snow-cover, each shift reflected in substantial ecological changes. The circumpolar tundra and desert were reduced several times to refugia as forests expanded into them, and reconstituted into continuous zones as the forests retreated (Lindsey, 1981). Known refugia during the last glacial maximum included central eastern Siberia, the Bering Strait area, central Alaska and Yukon, Kodiak Island, and parts of Banks, Baffin and Ellesmere Islands, west Greenland, Iceland, Svalbard and Norway (Haber, 1986).

As ice was shed from the land, sea level rose. Relieved of its weight of ice the land too has risen, but many former coastal areas, including sites of human settlement, remain under the sea around present-day northern coasts.

Southern deglaciation. Possibly because of its immensely larger bulk, antarctic ice has changed far less during the same period. The ice mantle has for long overlapped the edges of the continent, unable to grow bigger without breaking off. The current annual increment of about 2000 billion tonnes of ice, due to snowfall, is lost each year by evaporation, melting and formation of icebergs at the rim. Climatic changes bringing more snow would result in more icebergs to maintain the balance; less snow would cause a reduction of the ice-sheet and a greater exposure of bedrock. There is evidence at many points around the continent that more rock is exposed today than in the immediately recent past, and that there were several retreats during the present interglacial (Péwé, 1960), though not on the scale of the northern retreat.

There is evidence too that ice near the coast has in the past been thicker, for many coastal islands and mountains carry the scars of glacial movement on their exposed surfaces. This probably resulted from the fall in sea level during the glacial period, which would have exposed more land around the continent, allowing the mantle to spread over a wider continent, to cover what is now shallow sea-bed, and to thicken inland along the line of the present coast. There have certainly been sea level changes since then, leaving raised beaches at many points along the exposed coasts of both continent and islands. On these are found most of Antarctica's meagre soils, plant communities and penguin colonies.

Ice has been shed also from South Georgia, Bouvetøya, Heard Island, Iles Kerguelen, Macquarie Island and other islands on the polar fringe, a process continuing into the present day on several of them, possibly all (Allison and Keage, 1986).

3.2.2 *Soil-forming processes*

Soils begin with the mechanical and chemical destruction of bedrock. The simplest ahumic soils are little more than fine rock debris sorted into horizontal layers by wind, water and chemical action, and holding moisture in their interstices. In humic or organic soils bacteria, algae and other microorganisms have colonized the debris, circulated and exchanged minerals in solution and evolved a complex organic chemistry. This process of maturation from ahumic to organic soils takes hundreds or thousands of years; time is an important component (Ugolini, 1986). Polar soils and their main characteristics are summarized in Table 3.1, and Figure 3.2 outlines stages in their development.

Cold and aridity. Cold inhibits or prevents plant growth and slows down chemical processes in soils, prolonging the period of maturation. Aridity too puts a brake on, for water is the medium for all the chemical reactions in both plants and soil, and these cannot happen in its absence. Chronically dry soils lack organic buffers and are often strongly sodic, saline, acid or alkaline, with free soluble salts that make them unattractive for plant colonization. Too little water brings minerals to the surface and leaves them there in an inhibiting layer; too much water may leach them away altogether, though that is seldom a problem in polar soils. Where there is adequate water present during the spring and summer period of plant activity, soils stand a chance of attracting living matter and starting the processes that ultimately mature them. Not surprisingly, however, a high proportion of polar soils are ahumic.

Permafrost and ground ice. Where the ground is frozen for most of the year,

Table 3.1 Polar soils: their characteristics and occurrence in polar regions.

Soil type	Characteristics	Occurrence
AHUMIC SOILS (less than 1% organic content)		
Frigic or Cold desert soils	Arid. Unsorted polygons with wedges. Desert varnish pavement and surface efflorescence. Chemical weathering minimal.	Dry high-polar areas north and south; restricted to mountains in sub-polar areas.
	Dry. Horizons develop; soluble minerals remain unleached	
	Moist. Sorted polygons, stone rings, fine earth spots. Chemical weathering; clays may be present. Horizons vage or absent; soluble minerals leached	Rare in high-polar deserts; prevalent in damp coastal areas e.g. of maritime Antarctic, and northern polar coasts.
Evaporites	Dry. High content of crystalline soluble salts.	Widespread in dry areas of both polar regions.
ORGANIC SOILS (more than 1% organic content)		
Lichen- or moss-covered soils	Moist. Low organic content (1–10%). Chemical weathering present; limited humic activity with staining. Associated superficial moss or lichen mats.	Most widely-spread polar vegetation in moist areas that are snow-free in summer.
Moss peat soils	Moist. High fibrous organic content. pH 4.0–5.0; often high nutrient content. Underlying mineral soil not integrated. Low decomposition rate.	Rare in high polar regions; characteristic of damper areas, e.g. of maritime Antarctic and Arctic fringe.
Ornithogenic and wallow soils	Moist. High organic content from bird or seal refuse. High nutrient levels, often toxic.	Widespread in coastal areas where birds and seals congregate, and on inland nunataks and bird-breeding cliffs.
Brown soils	Moist High organic content (more than 10%). Associated with vascular plants. Humus integrated with mineral content. Often high rate of decomposition	Widespread in Arctic where moisture permits; characteristic of richer tundra. Almost unknown on continental Antarctica, widespread in patches in maritime Antarctic.

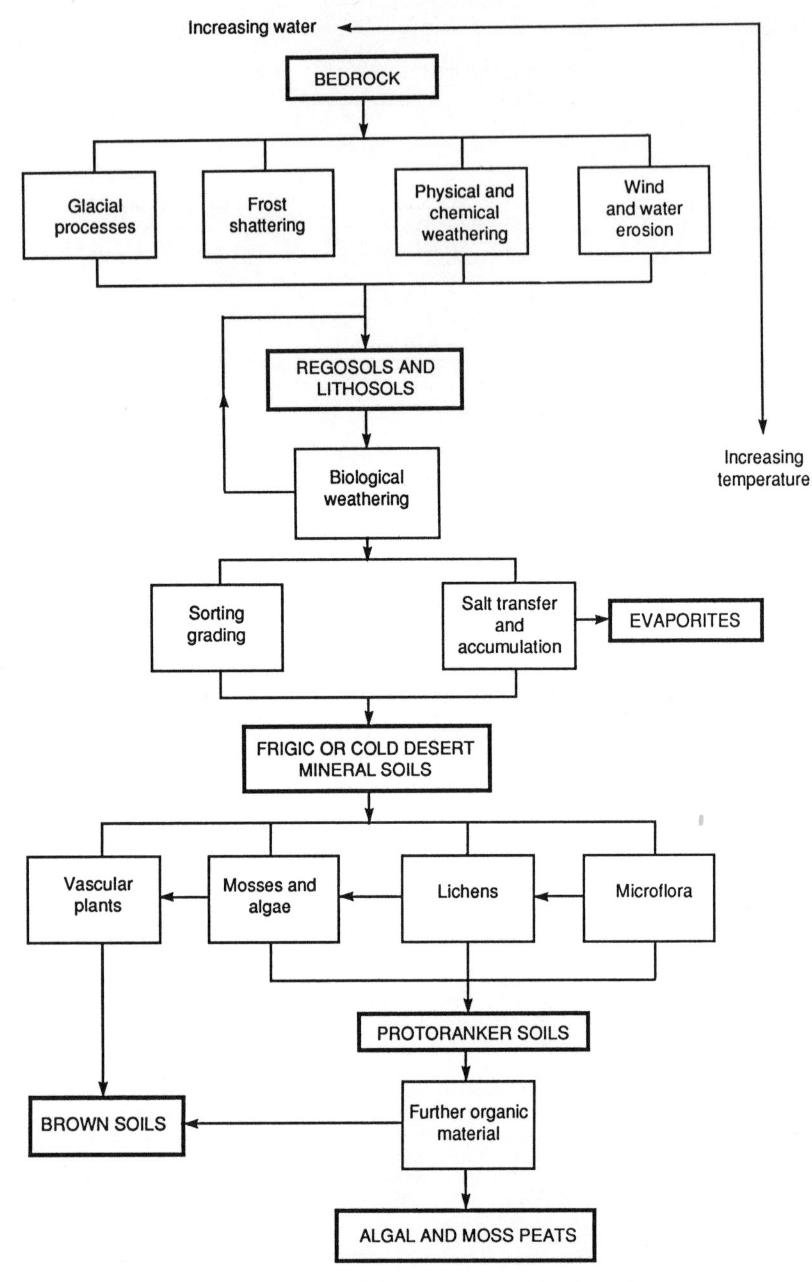

Figure 3.2 Stages in the development of polar soils.

soils that contain water remain solid except for a brief spell each summer, when they thaw to a depth that varies with latitude and local conditions. The thaw does not usually exceed about 1 m on flat ground, but on south-facing slopes it may penetrate to 1.5 m (Chernov, 1985). Most soil-forming processes are restricted to this surface active layer (Figure 3.3). Recent drilling for oil and natural gas has shown that continuous permafrost underlies much of the Arctic, reaching thicknesses of 600–1000 m in the coldest regions and extending out under the sea in the Arctic basin (Figure 3.4). It is widespread in the Antarctic, and discontinuous or patchy permafrost underlies much of the Subarctic (Washburn, 1979).

Permafrost restricts drainage; melt-water cannot sink into the ground and in summer the thin active layer soon becomes waterlogged. The ice forms a mechanical barrier to roots, limiting the penetration of shrubs and restricting the growth, form and stability of trees. Surface snow and vegetation help to insulate the ground in spring, restricting the depth of thawing. Local thicknesses and effects of permafrost are discussed by Tedrow and Brown (1967) for the North American high arctic tundra, Chernov (1985) for the Siberian Arctic, and by papers in Campbell (1966) for Antarctica.

Bedrock, lithosols and regosols. Mechanical weathering is rapid in cold desert regions. Polar deserts are characteristically littered with rock rubble, frost-shattered from bedrock on the spot (lithosols) or redistributed, ground and compacted into pavements by glacial action, and re-exposed after the retreat of the ice sheets (regosols). Several kinds of weathering are evident (Figure 3.5). As in all deserts, rapid changes of surface temperature are an important disruptive force. In McMurdo Sound, Antarctica, Tedrow and Ugolini (1966) and Tedrow (1977) estimated that shifts of 30–40° C occurred several times daily in the surface of rocks under summer sunshine,

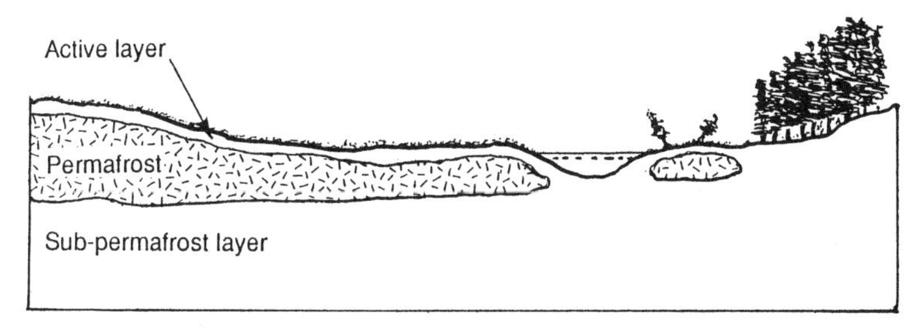

Figure 3.3 Permafrost is continuous under the open tundra (left) but not under the lake or forest (right). Trees growing where permafrost is present tend to be stunted and blown over by storms.

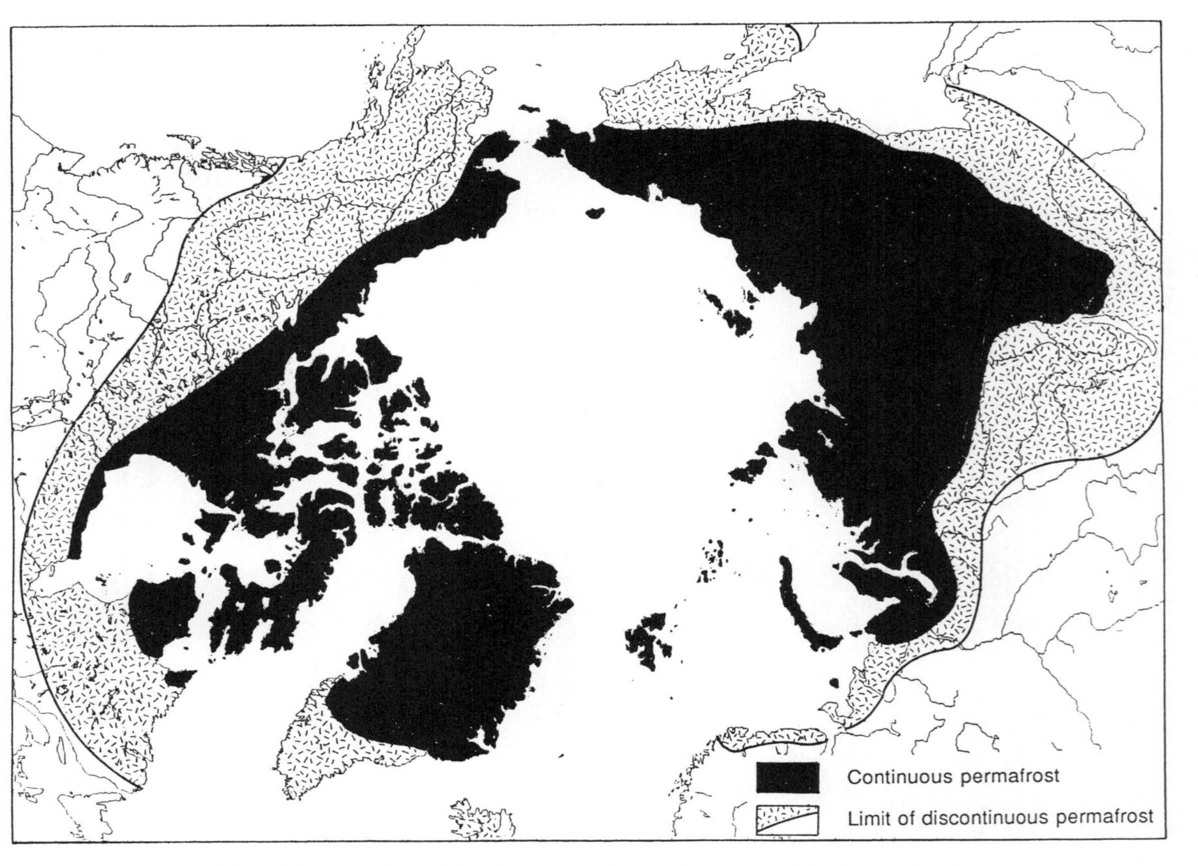

Figure 3.4 Distribution of continuous and discontinuous permafrost in the Arctic.

Figure 3.5 Exfoliation of rocks due to frost action. Photo: Guy Mannering.

and Nichols and Ball (1964) found bare ahumic soils passing through the freeze-thaw cycle over a hundred times in the course of a summer. Salt solutions penetrate cracks and exfoliate rocks on evaporating (Wellman and Wilson, 1965). Wind-borne grit and ice erode antarctic basalt boulders at a rate of 1.5 mm per year (Evteev, 1964), producing pavements of consolidated rock rubble with each fragment cut to a tricorn and wind-polished (Nichols, 1966). Chemical weathering is rare in cold desert soils, though limonite, clay minerals and other typical products have been reported, and the breakdown of ferromagnesian minerals to reddish iron oxides is apparent on many rock pavement surfaces (Tedrow and Ugolini, 1966; Allen and Northover, 1967; Ugolini, 1970).

Crustose lichens. With only limited capacity for retaining water, bedrock, regosols and lithosols are poor material for plant colonization. Completely dry inland regions where they predominate, for example the so-called 'oases', support little vegetation. Where there is a hint of moisture, rocks that have stable, non-foliating surfaces and are sheltered from wind-blast support crustose (encrusting) lichens, the characteristic vegetation of polar deserts. These cling without roots and survive on atmospheric moisture, trickles of melt-water or trapped snow and rain. Sometimes colourful with red or orange pigments, they occur on the dry rocky plains of northern Greenland, Alaska and Canada, on the Siberian islands, on dry rocks among the tundra where nothing else grows (Richardson and Finegan, 1977), and throughout Antarctica, including inland mountains less than 500 km from the South Pole (Greene and others, 1967, Cameron *et al.*, 1971). Crustose lichens are an additional weathering agent, tending to penetrate and defoliate rock surfaces (Walton, 1985). In favourable situations they form nuclei for small communities of microflora and fauna, their debris accumulating as an organic substrate that fosters mosses and angiosperms.

3.2.3 *Ahumic soils*

True soils develop from these unpromising beginnings when fine materials, accumulated by water or wind, remain stable long enough to differentiate chemically into horizontal bands or horizons (Campbell and Claridge, 1987).

Cold desert soils. In dry polar regions, where ionic mobility is low, this is a long, slow process; the cold desert or frigic soils (Campbell and Claridge, 1969) formed under these extreme conditions have been described only in Antarctica, the driest regions of the Arctic and the cold mountain deserts of Tien Shan (Tedrow and Ugolini, 1966). Usually wind-blown, they fill pockets among coarser rubble, or on a larger scale cover many hectares of flat ground to depths of half a metre or more, firm and sometimes consolidated. If soft enough to show footprints and vehicle tracks, they retain them for many years.

Typical cold desert soils have three layers (Figure 3.6A), of which the lowest is permanently frozen sand or gravel. They tend to be alkaline, with free carbonate and calcium, potassium, sodium, magnesium and other ions distributed in patterns that vary with age and the amount of leaching. In Antarctica, Campbell and Claridge (1969) distinguish three categories—ultraxerous, xerous and subxerous—respectively from the very dry interior, the dry coasts and the damper coasts, recognizable by the amount of ionic movement that has occurred.

Salt concentrations near the surface may make them especially inhospitable to vegetation; only the soils of damper areas, in which salts are distributed evenly or concentrated at depth, are likely to support plant propagules. However, ahumic soils are seldom completely abiotic. Benoit and Hall (1970) and Cameron *et al.* (1970), working in the Dry Valleys oasis area of McMurdo Sound (and using techniques that would later be used for detecting life on Mars), reported small populations of bacteria in samples taken even from the driest and least promising areas. Similar microfloras have been found almost everywhere in both polar regions where ahumic soils exist.

Volcanic vent soils. Special soil-forming conditions occur close to warm volcanic vents. In Antarctica heat and moisture from steam and melting snow near Mount Melbourne, Mount Erebus, and on volcanic islands in the South Sandwich and South Shetland groups, cause intense chemical weathering, with the production of clay minerals, free chloride and sulphate radicals, and strongly acidic soils. The warm, permanently damp clay provides a localized substrate where microorganisms and moss mats flourish (Broady *et al.*, 1987; Ugolini, 1966, 1967; Vincent, 1988).

Evaporites and saline soils. Evaporites are ahumic soils with a high proportion of soluble crystalline minerals, formed where drainage water carrying dissolved minerals has accumulated and evaporated. Normally associated with hot deserts, they are not uncommon in dry polar regions, surrounding saline lakes (Ugolini and Tedrow 1966). Saline soils form near the sea when salts are blown inland from spray in summer, or in salt-laden snow lifted from sea ice in winter. Sea spray may contain up to 25% of solid materials, organic and inorganic, and contributes significant amounts of nitrate and organic debris to coastal soils. On Signy Island sodium ions may release potassium from soil solution, making available an important plant nutrient (Allen and Northover, 1967). In temperate regions they support specially-adapted halophilic (salt-loving) species of plants; in polar regions they tend to be barren. Chemical characteristics of some Signy Island soils appear in Table 3.2.

3.2.4 *Organic soils*

Protorankers, rankers and moss peat. In dry polar deserts ahumic soils tend to remain ahumic indefinitely. Only in warmer, damper regions have they a

Table 3.2 Chemical characteristics of some Signy Island soils: dry weights. From Allen and Northover (1967) and Northover and Allen (1967).

	pH	Organic (%)	CN	P	Exchangeable cations (meq 100 g) Ca	Mg	K
Ahumic soils							
Schist gravel	6.1	2.0	0.04	0.16	13	5	3
Schist clay	6.3	2.5	0.06	0.14	30	10	8
Marble debris	8.4	2.2	0.05	0.15	660	15	3
Humic soils							
Moss peat	4.6	94	1.1	0.09	167	15	3
Brown soil	5.4	11.8	0.57	0.28	76	40	18
Brown mineral soil	5.4	38	1.3	0.22	193	208	110

chance to acquire the organic debris—mostly fragments of algae, fungi, lichens and mosses—that allows them to develop further into humic soils (Figure 3.6B–D). The most advanced northern soils are arctic brown soils, which form on damp but well-drained open slopes colonized by flowering plants. Their colour and loamy consistency indicates that organic solutions are working among the soil minerals. Also showing a degree of maturity are ranker soils (Kubiéna, 1953; Tedrow, 1977)—coarse mats of vegetation several centimetres thick overlying and to some degree penetrating mineral soils below. The closest polar desert approximations are protoranker soils—moist soils of very limited organic content underlying patches of moss and lichen (Figure 3.6C).

Protorankers occur widely in coastal Antarctica, wherever there is moisture enough in summer to support vegetation mats (Tedrow and

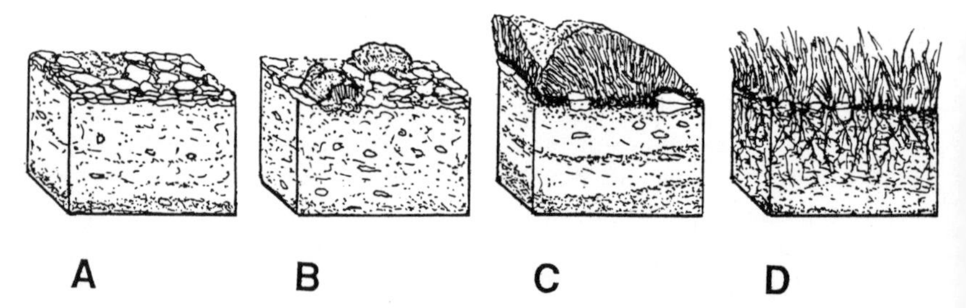

Figure 3.6 Polar soils. Frigic soils (*A*) typically have a surface of wind-polished surface pebbles overlying fine materials; permafrost may be present at 15–20 cm. Small colonies of moss and lichen (*B*) cause only slight changes in structure; in protorankers (*C*) larger clumps form superficial peaty masses, produce some mineralization and raise the permafrost level. Roots of grasses and other vascular plants (*D*) give rise to brown soils.

Ugolini, 1966; Ugolini, 1977; MacNamara, 1969). They tend to be alkaline, with soluble salts at the plant–soil interface and sometimes a brown mineral horizon below. Theoretically they mark a stage in the formation of organic soils, but as neither mosses nor lichens put down penetrating roots, and earthworms are lacking, organic material remains mostly superficial. Organic content of the soil beneath ranges from 3.1% under lichens and 5.3% under mosses (Rudolph, 1966) to approaching 10% under lichen mats (Glasovskaya, 1958). The vegetation cover has an insulating effect in spring, keeping the soil cool and humid and the level of permafrost high (McCraw, 1960).

In maritime Antarctica some Signy Island soils underlying moss peats (Allen and Northover, 1967; Holdgate *et al.*, 1967), and at least two of eight soil groups identified by Everett (1969) from Trinity Peninsula, seem likely to be protorankers. Where moisture is limited and productivity low the moss covering remains thin. Where productivity is higher the carpet thickens and becomes peaty; Allen and Heal (1970) reported moss peat up to 2 m thick, overlying rather than mixing with the mineral soil beneath.

Brown soils. True organic soils require the presence of vascular plants with penetrating roots, and a lively population of soil flora and fauna, preferably including earthworms. On the arctic tundra Tedrow (1977) distinguishes arctic brown soils on well-drained sites, and tundra soils on wetter areas, the latter grading into waterlogged bog soils. These are poorly developed brown soils, low in organic content though overlain by mats of plant material, a strongly mineralized subsurface layer and a permafrost base. They often carry an active microbiota of bacteria, algae, nematodes and other creatures that mediate slow decomposition and the incorporation of organic materials into the soil (Figure 3.6D).

Brown soils of this level of maturity are found as far north as Devon Island (75°30′N) in the Canadian Arctic, forming mosaics with polar desert soils (Walker and Peters, 1977). In Siberia they are more widespread and varied than in North America, where glaciation was heavier and longer-lasting. Simple brown soils occur in the maritime Antarctic, on damp north-facing slopes in association with Antarctica's two species of flowering plants (Allen and Heal, 1970).

Richer, more mature brown soils occur in polar fringe regions of both hemispheres, usually underlying grass and herb meadows. Though the soil may be frozen for part of the year, root growth is vigorous, earthworms may be active, there is plenty of organic debris to be rotted down, and the decomposing agents are more abundant than in the dry deserts.

Figure 3.7 Ornithogenic soils. A penguin colony covering many hectares (above) accumulates guano, especially around the nesting areas (below). Photo: John Darby.

Ornithogenic soils. Soils saturated by concentrated bird droppings occur throughout the world; those of seabirds in dry areas are of particular economic interest as origins of phosphate deposits (Hutchinson, 1950). They are found at both ends of the world (Syroechkovskiy, 1959), being especially prominent at penguin breeding colonies in coastal Antarctica. Colony sites include massive quantities of highly phosphatic and nitrogenous guano, and are scattered also with feathers, food, egg membranes and the carcasses of unsuccessful chicks (Figure 3.7).

On continental Antarctic coasts these deposits weather slowly (Campbell and Claridge, 1966; Orchard and Corderoy, 1983; Ugolini, 1972), with little modification of underlying mineral soils. At old abandoned sites friable, loamy subsoil of silt and clay develops, and at long-abandoned sites most soluble salts gradually disappear. Some have enhanced populations of bacteria (Boyd and Boyd, 1963; Vincent, 1988). Though these might seem ideal sites for vegetation, they are not especially attractive to algae, lichens and mosses.

In the warmer, damper conditions of maritime Antarctica, colony sites examined by Myrcha *et al.* (1985) showed more active leaching and mineralization, notably phosphatization, of the underlying soil, with rapid run-off of soluble nutrients to the sea. Nunataks and cliffs with small local populations of gulls, petrels and skuas gather deposits which seem more encouraging to plant growth (Siple, 1938; Lovenskiold, 1960; Rudolph, 1966), perhaps because concentrations of guano are less overwhelming.

3.2.5 *Patterned ground*

Bare or thinly-covered polar soils often show characteristic hillocks, or patterns of circles, rectangles, polygons and stripes; these may be marked out in shallow furrows and ridges, or by lines of stones, on scales of 1 m to 30 m or more (Tedrow, 1977). The patterns are common to both hemispheres and appear only on open ground, not under permanent ice (except for some 'fossil' patterns representing an earlier interglacial), or under long-standing snow drifts or thick vegetation. Depending on the presence of soil moisture, they are absent from very dry ground and vary in structure according to wetness.

On moderately dry ground, for example along the coast of Antarctica, unsorted polygons are convex shapes in the ground up to a few metres across, ringed by a network of furrows that mark the positions of persistent cracks (Figure 3.8). As temperatures fall with the approach of winter, the soil freezes and contracts, forming the cracks which open to admit snow and surface rubble. When the soil expands again in spring the contents

Figure 3.8(*A*) Patterned ground: unsorted (fracture) polygons in soft volcanic rock. Photo: Guy Mannering.

Figure 3.8(*B*) Solifluction lobe on moist slope showing partial sorting of coarser material to margins of the feature. Moss is predominantly *Drepanocladus uncinatus*. Signy Is. Photo: Ron Lewis Smith.

of the cracks are compressed into wedges. The cracks recur each year in the same places, the wedges thicken annually, and the ground between buckles into irregular domes, ridges, or pancakes with raised edges. Parallel cracks form where the soil is patchy or irregular, or the ground sloping; on flat ground lesser cracks appear at angles to the main system, forming irregular polygons. The tops of the wedges are often covered with gravel and may be awash during the spring thaw.

Sorted polygons occur on wetter ground; they are common for example on bare ground in maritime Antarctica and widespread on the arctic tundra (Figure 3.9). The simplest are fine earth spots, which stand out like molehills on a lawn and are often defined by rings of large pebbles or stones, which they appear to have shouldered aside. During freezing in wet soil, water is drawn into horizons which solidify in parallel layers, forcing soil and stones upward. On thawing in spring the fine soil collapses and falls past the stones, which thus become isolated at the surface and encircle the actively freezing and thawing material. On hillsides, stone

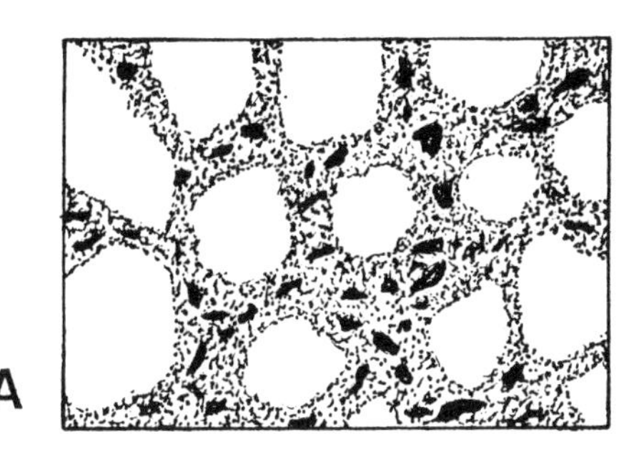

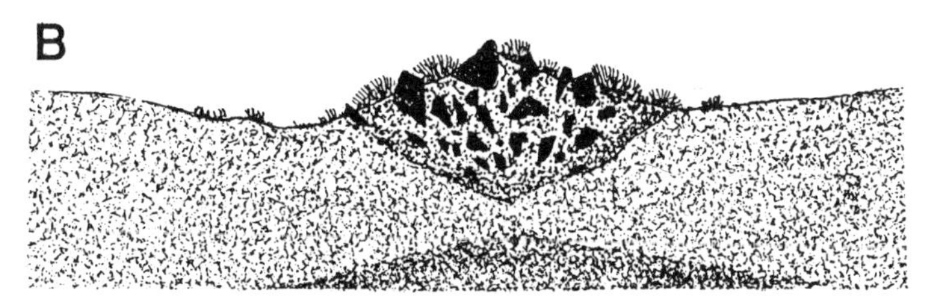

Figure 3.9 Patterned ground: sorted polygons in moist soil. (*A*) Polygons 1–3 m in diameter, of fine particles stirred and kept bare by soil movement, are bounded by (*B*) more stable ridges of stones and well-drained soil, which support a heavy growth of moss; the permafrost rises beneath the ridges. After Wilson (1952).

stripes form in much the same way across the slope (Chambers, 1966a, b, 1967; Holdgate *et al.* 1967).

On a larger scale, in very wet ground, lenses of ice several metres thick may accumulate over a number of years, causing hummocky ground; on thawing they create pits from which originate many tundra ponds. Very large lenses up to 50 m thick form extensions of the permafrost, giving rise to steep-sided hills called pingos which in many tundra regions provide the highest relief. Pingos may be many centuries old and covered with permanent vegetation.

Though summer moisture and warmth encourage vegetation, soil mobility discourages it, and the most active spots in patterned ground remain uncolonized (Figure 3.9). Stable furrows, ridges and hillocks provide a range of environmental conditions which plants select according to their needs. Patterned ground is thus reflected in a patterned carpet of mosses and flowering plants. Patterning is widespread and often spectacular in the Arctic. Though widely reported from Antarctica, it is seldom so striking in the drier southern soils.

3.3 Land plant communities

Northern and southern polar plant communities have few species in common, but fall readily into similar biogeographical categories according to cold and aridity. Several workers have devised systems of classifying either habitat or vegetation for one or both regions. For the north see for example Polunin (1951), Young (1971) and Bliss (1981); Bliss (1981–9) provides a table of comparison between eight published biogeographical systems. For the south see Bliss (1979), Greene (1964), Holdgate (1977), Stonehouse (1972, 1982), Pickard and Seppelt (1984) and Smith (1984). Among the few who have tried to compare the two regions, French and Smith (1985) are concerned particularly with subpolar tundra, and Aleksandrova (1980) prescribes a simple geobotanical regional zoning that covers both polar and subpolar zones at both ends of the world. Here I use Aleksandrova's system for the north, and a compatible though slightly different system (Stonehouse, 1982) for the south. These are summarized in Tables 3.3 and 3.4.

3.3.1 *Northern plant communities*

Polar desert and tundra. Tundra is a Lappish word, probably derived from the Finnish 'tunturi', and meaning 'treeless heights'. It was originally

Table 3.3 Zonation of vegetation in the circumpolar Arctic; after Aleksandrova (1980).

Region	Subregion	Characteristics	Boundary
Polar desert	Northern belt	Oceanic species of crustose lichens dominant. Fewer than 50 angiosperm spp.	Northern limit of land
	Southern belt	Continental species of crustose lichens dominant. Up to 67 angiosperm spp.	Line between exteme northern and continental islands
Tundra	Arctic tundra	Arctic-alpine and arctic dwarf shrubs: *Salix polaris*, *Dryas integrifolia*, *Cassiope tetragona* etc.	2°C July isotherm
	Subarctic tundra	Semi-prostrate hyparctic shrubs: *Betula nana*, *B. exilis*, *B. glandulosa* shrubs, etc. and dwarf shrubs, *Vaccinium* spp., *Empetrum hermaphroditum*; *Ledum decumbens* etc.	6°C July isotherm Forest limit

applied to the open plains of northern Kola Peninsula (Kol'skiy Poluostrov). Now it is used widely in a geographical sense for all the plains, vegetated but treeless, that lie north of the boreal forest in Europe, Siberia, North America and Greenland, up to the boundary of the polar desert. Polar desert is the area of bare rocks and soils beyond the tundra, where only a thin, patchy covering of plant life is found.

Confusingly, the same terms are used also for the plant communities of these regions. In its botanical sense tundra is a working abbreviation for arctic tundra vegetation, the tough, low-lying vegetation that typifies the arctic treeless area. There is also mountain tundra, a similar kind of vegetation growing on high mountain slopes in lower latitudes. Similarly the term polar desert covers both the area and the vegetation characteristic of it.

Table 3.4 Zonation of vegetation in the circumpolar Antarctic (after Stonehouse 1982).

Region	Subregion	Subdivisions	Characteristics	Boundary
Antarctic	Continental Antarctica	East and West Antarctica, Alexander Land, Charcot Land, islands close to the continent: three ecological zones—High plateau, Inland slopes, Coast.	Mostly ice-covered; mean temperatures of warmest month below 0° C. Annual precipitation < 20 cm; clear skies inland. Ahumic and protoranker soils; algae, lichens, mosses; poor soil fauna of nematodes, mites, collembolans etc.	Extends to pole
	Maritime Antarctica	Antarctic Peninsula south to 69° S on west coast, 64°S on east coast, and neighbouring islands: South Orkney, South Shetland and Sandwich Is. Peter I Øy also (probably) Scott I. and Balleny Is.	Mostly ice-covered; mean temperatures of warmest months −1° to +2°C, winter means rarely below −15° C. Annual precipitation 20–100 cm; cloudy. Ahumic and protoranker soils; algae, lichens, mosses, 2 spp. of angiosperms, poor soil fauna of nematodes, mites, collembolans etc.	0° C July isotherm
	Periantarctic islands	South Georgia, Bouvetøya, Heard I., Macdonald Is.	Glaciated uplands (not Macdonald Is.); mean temperature of warmest months 0° to +6°C, winter means rarely below −2°C. Mostly brown soils; mosses, ferns, many angiosperm species, soil fauna includes earthworms, beetles.	Northern limit of pack ice Antarctic Convergence

Arctic desert vegetation. The northern polar desert has few macroscopic plants or none; where present the vegetation forms a single thin layer, with cryptogams (algae, lichens, mosses and hepatics) dominant. Mosses and lichens provide much of the patchy ground cover, forming mats rather than carpets with bare ground between. Angiosperms (flowering plants) are present but insignificant. Lichens, particularly encrusting forms, are the most prominent plants, dominant in many communities and providing the highest biomass. *Neuropogon sulphureus* (= *Usnea sulphurea*), a bipolar species, is typical, growing in colourful patches among species of the genera *Collema*, *Ochrolechia*, *Pertusaria*, *Toninia* and others.

Algae, especially blue-green forms, occur among the lichens. Mosses are represented mainly by small clusters including species of *Bryum*, *Pohlia* and *Myurella*; *Rhacomitrium lanuginosum* and *Andreaea* spp. are common on poor soils and *Onchophorus wahlenbergii* grows in wetter areas; some larger tundra mosses are present but comparatively rare. Liverworts grow chiefly in association with mosses, sometimes forming black incrustations.

Angiosperms are usually tiny, growing in isolated clusters and nowhere forming the bulk of the cover. Only about 60 species thrive in the low summer temperatures. Their most characteristic form is compact cushions (for example poppies *Papaver*, moss-campions *Draba* and saxifrages *Saxifraga*), or small tufts (grasses *Phippsia algidae* and *Poa abbreviata*). In their horizontal distribution plants often follow soil patterning (p. 77), mosses, lichens and angiosperms aligning themselves along the cracks between polygons, with crustose lichens occupying the centres.

Notably absent from this zone are the sedges *Carex stans*, and the angiosperm families Polypodiaceae, Liliaceae, Betulaceae, Empetraceae and Vacciniacae, all well represented further south. Sphagnum bogs and peat mires are missing; occasionally the ground is wet enough to support moss mires, involving ahumic soils and an overgrowth of mosses *Orthothecium chryseum*, or *Campylium* and *Bryum* spp. forming protoranker mats on top. Among lichens the genus *Cladonia* is very rare, except for *C. pyxidata* and other cup-like forms.

Within the arctic desert region Aleksandrova distinguishes northern and southern subzones. The northern belt includes Zemlya Frantsa-Iosepha, the northernmost islands of Severnaya Zemlya, the De Long Islands, the northernmost group of Novosibirskiye Ostrova, and the coldest, driest, most completely icebound islands of the Canadian arctic archipelago (Borden, northern Prince Patrick, Lougheed, Ellef Ringness, Amund Ringness and Meighen Islands). The southern belt includes the northern tip of Novaya Zemlya, the bulk of Severnaya Zemlya, Cape Chelyuskin, several of the less extreme Canadian arctic islands, the northern coasts of Devon, Axel Heiberg and Ellesmere Islands, northern Greenland, and Nordaustlandet and Kong Karls Land in Svalbard.

While the northern belt has an almost exclusively polar desert flora, the southern belt has climatically-favoured areas where tundra species occur patchily, and forms a transition zone between high arctic and tundra. Its flora includes a few vascular cryptogams (ferns). The boundary between the edge of the southern zone and the tundra is the 2°C July isotherm. Longitudinally the whole desert region is divided into Barents, Siberian and Canadian provinces, with similar but distinctive plant assemblies.

Arctic tundra vegetation. Arctic tundra vegetation is a mosaic of plant communities, usually compact, wind-sculptured, and less than 1 m high. Lichens and mosses are often prominent, but tundra communities also include shrubs, sedges, grasses and forbs (flowering herbs other than grasses). Composition of the communities varies according to soil, aspect in relation to sun, drainage, length of snow-cover and other variables. At its richest, on good soils in warm, sheltered spots, tundra can form lush meadows, thickets of tussock grasses and low shrubs. At its poorest, on rocky soils in harsh, unsheltered environments, it is a thin layer of vegetation, including flowering plants, that seems to survive only with difficulty.

Tundra vegetation forms a semi-continuous zone all the way round the Arctic basin, wherever there is land far enough north to support it. In northern Scandinavia there is only a narrow band a few kilometres wide along the northeastern coast of Kola Peninsula. In central Siberia, Alaska and Canada tundra forms a belt several hundred kilometres wide on the mainland and extends far out onto the islands beyond the continental coasts. Two subzones are usually distinguished. The southernmost subzone is one of transition from forest to typical tundra. Where the treeline is sharply defined it may hardly exist at all; more often it is a narrow band of forest-tundra, in which stands of trees alternate with dense shrubby tundra. The northern zone is one of progressive impoverishment, where tundra vegetation thins rapidly in height, density and number of species. Ground with less than about 50% cover is called fellfield; below about 20% fellfield becomes polar desert.

Typical tundra forms a thin, semi-complete vegetation covering in which grasses or low shrubs up to about 50 cm tall are dominant; spaces within it are caused by bare rock outcrops, and characteristic patches of mobile soil which all plants find difficulty in colonizing. Many tundra species occur in polar desert and forest as well, but only in tundra do they dominate their communities. Tundra vegetation includes a greater variety and profusion of algae, lichens, liverworts and mosses than polar deserts, but a more striking and diagnostic difference is the variety and form of angiosperms, notably low shrubs, herbs and grasses.

Aleksandrova rather confusingly calls the northern and southern belts arctic tundra and subarctic tundra, separating them by the 6°C July isotherm. In the narrower northern subregion the richest plant assemblies are dominated by dwarf shrubs, for example polar willows *Salix polaris*, the avens *Dryas octopetala*, *D. punctata* and *D. integrifolia* and the heather *Cassiope tetragona*; these grow at most only a few centimetres high, forming thin stands that barely cover the ground. This region has the longer history of continuous occupation by tundra, for at no time during the post-glacial period was it invaded by forest.

In the southern zone, where the growing season is longer, thickets of larger, denser shrubs become possible. Characteristic plants include birches *Betula nana*, *B. exilis* and *B. glandulosa*, bilberries *Vaccinium uliginosum* and crowberries *Empetrum hermaphroditum*, which grow recumbent or semi-prostrate, in some patches forming a complete ground cover. Along its southern edge stands of low trees intermix with the shrubs, in places grading insensibly into forest.

These communities dominated by dwarf and recumbent shrubs represent the best conditions of soil, warmth and moisture in their respective subregions. Other communities, less diagnostic but equally representing the tundra, occur in less optimal conditions. Aleksandrova divides each tundra subregion latitudinally into northern and southern zones, providing four belts of progressive floristic enrichment and diversity within the tundra biome. The northernmost zone has the least continuous plant cover, the most patterned ground exposed, and the narrowest range of communities. The southernmost has the densest vegetation, the most complete cover, and the widest variety of both angiosperms and plant communities.

Tundra ecologists stress the mosaic characteristic of this vegetation; many different kinds of tundra communities are reviewed in individual articles in Bliss *et al.* (1981), which cover the whole arctic area. For detailed surveys of lowland tundra on Devon Island, northern Canada, see Bliss (1977); for a wide-ranging though equally detailed account of Soviet tundra see Chernov (1985), and for a most readable though more general survey of tundra vegetation see Sage (1986). A simple approach to the diversity of tundra communities is to classify them as dry, moist or wet, according to dampness underfoot.

Dry: fellfields and barren grounds. Widespread in the north, and restricted to dry uplands in the south, are expanses of rough, open terrain with virtually ahumic soils. For much of the year they appear barren; with summer moisture the dried-out cushions and patches of vegetation burst into life, yielding arctic poppies, avens, moss campions, saxifrages—flowers familiar from the arctic desert, here abundant enough to provide a scattering of colourful, threadbare mats. Adding to the textures are chickweeds *Stellaria*, wintergreen *Pyrola grandiflora*, willow-herb *Epilobium latifolium*, mountain vetch *Astragalus alpinus*, dwarf willows, heathers, Labrador tea *Ledum decumbens* and other small woody shrubs, alternating with thin grasses, mosses and lichens.

Moist: shrub and heath lands. The central tundra belts are steppe-like plains where low shrubs, sedges and grasses form a semi-continuous cover of vegetation up to 1 m high. In moist lowlands a dense ground cover of short grasses, mosses, lichens and occasional ferns (*Woodsia* spp.) grows from peaty, sometimes waterlogged soils. Dwarf willows, birches and alders (*Alnus*) form the emergent vegetation. Better-drained uplands have thinner soils and a ground cover of grey-green lichens, notably *Cladonia* spp., the so-called reindeer mosses. The tundra is widely grazed by mammals, especially voles and lemmings that burrow in the undergrowth.

Small shrubs—cranberries *Vaccinium vitis-idaea*, bilberries, bearberries *Arctostaphylos* spp., crowberries, Labrador tea, and arctic heather—grow in patches; lupins *Lupinus arcticus*, buttercups *Ranunculus lapponicus*, windflowers *Anemone parviflora* and louseworts *Pedicularis* spp. produce colourful flowers in summer, and several of the shrubs produce bright red leaves and berries in autumn. Where snow persists late in summer and the growing season is short, meadows of short grasses and herbs appear among the shrubs.

Wet: mires, marshes and ponds. Undulating coastal plains and other ungraded lowlands, underlain by permafrost, in summer form some of the tundra's wettest areas. Ponds, marshes and mires fill the hollows, alternating with higher ground which emerges first from under the snow and dries out in the course of summer. From wet, acid, peaty soils with partly-decomposed debris emerge tussock grasses, rich swards of pendent grass *Arctophila fulva* and tundra grass *Dupontia fischeri*, sedges *Eriophorum* and *Carex*, rushes *Juncus* spp., cotton grasses *Eriphorium angustifolium* and dark green, hummock-forming mosses. Shrubby alders and willows, buttercups *Ranunculus pallasii* and marsh marigolds *Caltha natans* line the pond edges, with grass swards that become favoured grazing grounds for aquatic birds. Ponds of the southern and central tundra warm quickly once their ice has melted, and build up a moderately rich flora and fauna.

The tundra–forest boundary. The treeline defining the southern tundra edge is often fringed by a forest-tundra zone, narrow in North America but broadening in Eurasia to a band up to 300 km wide. Chernov (1985) describes a transition zone of multi-layered vegetation, rich in species, patchy because of local variations in drainage and microclimate. On dry ground thickets of willow, birch and alder alternate and mingle with moss and grass meadows; in the hollows are moss mires of *Sphagnum* and *Hypnum* spp., yielding to sedges, cotton grass, pendent grass and mare's tails *Hippuris vulgaris* spp. in the wetlands. South-facing slopes where snow lies late support meadows of grass and herbs; drier ridges carry a dense growth of dwarf shrubs. Shrubs that grow knee-high further north reach waist-height and above at the forest edge.

Isolated stands of small, wind-stunted trees are characteristic, in North America usually white or black spruce *Picea glauca*, *P. mariana*, in Eurasia Scots or stone pine *Pinus sylvestris*, *P. pumila*. Forming islands in the tundra, their height is determined by winter snow depth; deep snow protects the body of the tree, but exposed tips of leaders and lateral branches are nipped by winter winds. Seeds are set in good seasons, but few are viable, germination is low, and seedlings are vulnerable to predation. Poor regeneration from seed is an important factor that inhibits the spread of forest into the tundra (Tikhomirov, 1961).

3.3.2 *Southern plant communities*

Southern biomes. In Stonehouse's (1972, 1982) classification of southern biomes (Table 3.3) the antarctic region includes the whole of Antarctica north to the Antarctic Convergence. Within it lie circumpolar continental, maritime and periantarctic zones, the first bounded by the 0°C summer isotherm, the second by the northern limit of pack ice. Aleksandrova's (1980) bipolar classification is primarily geobotanical, though it recognizes similar boundaries and differs mainly in nomenclature. Her 'antarctic polar deserts' region covers the continent, peninsula and islands up to the northern limit of pack ice, and is divided into southern and northern subregions separated by the 0°C summer isotherm, which approximates to the isoline of zero radiation balance.

Aleksandrova's system thus includes Stonehouse's two southern zones in one region, her two subregions corresponding almost exactly to his continental and maritime zones. This classification emphasizes a legitimate interest of a bipolar geobotanist—the existence of a southern polar desert directly comparable to the better-known northern one. Stonehouse is concerned more to emphasize environmental and ecological differences between the continental and maritime regions, supporting Holdgate's (1964) original concept of a maritime antarctic province, and in accord with other biologists (e.g. Smith, 1984) whose main interests are antarctic.

Continental plant communities. First impressions in the field suggest that large areas of ice-free Antarctica are sterile. However, many of the seemingly-barren rocks carry a sparse flora of algae or lichens, and few damp crevices or patches of soil are without at least a microflora of algae, bacteria or fungi. Cameron (1972) cultured green or blue-green algae from some of the driest and least-promising looking areas of the McMurdo Dry Valleys oasis. At 3600 m on Mount Erebus, Ross Island, where no surface vegetation was apparent, Janetschek (1963) recorded bacteria, blue-green algae and microfungi a few centimetres down in the volcanically-warmed soil. From dry stone pavements in the Vestfold Hills oasis Broady (1986) reported green and blue-green algae, zoned in slightly damp subsurface soil under and around translucent stones, in areas apparently devoid of vegetation. Vincent (1988) reviews these communities thoroughly.

Algal, lichen and moss communities grow wherever there is a hint of moisture, in a variety of habitats all over the continent. On mountain flanks within 5° of the South Pole they cover less than 1% of the ground; on damper coastal flats and uplands they may spread to 15% cover or even more. Very few species are involved; just how few are present is hard to tell because collections are patchy and taxonomy confusing.

Floral compositions of some representative antarctic plant localities appear in Table 3.5, where they are compared with a single northern high tundra community. Smith (1984) credits continental Antarctica with 15 taxa (provisionally species) of mosses, 100 of lichens, one hepatic, with no ferns, pteridophytes or angiosperms. Following intensive studies of sites in Victoria Land, East Antarctica, Janetschek (1967) gave a provisional list of at least 150 species of algae, over 32 species of lichens and 8–10 species of mosses—a representative selection from a high-latitude desert area.

A comparable dry desert area of the arctic might yield a similar selection of cryptogams, but there would probably be up to a dozen species of flowering plants scattered among them. In comparable *latitudes* of the

Table 3.5 Numbers of native plant species in selected Antarctic and Arctic localities. Antarctic data from Smith (1984), Devon Island data from Bliss (1977).

Locality	Lichens	Mosses	Liverworts	Ferns	Angiosperms
Continental Antarctic	125	30	1	0	0
Antarctic Peninsula	100	45	5	0	2
Maritime Antarctic	150	75	25	0	2
South Georgia	160	175	85	7	19
Iles Kerguelen	120	85	45	8	22
Heard I.	52	16 +	?	0	8
Macquarie I.	55 +	75	60	5	34
Devon I. (Arctic)	182	132	30	9	90

arctic the flora is much richer (Table 3.4). On 43 km^2 of Truelove Lowland, Devon Island (75 °N), there are more species of lichens, liverworts, mosses and ferns, and some 90 species of angiosperms (Bliss 1977). The environments are similar; without doubt the single most important difference is the greater opportunity for recruitment from adjoining forest, steppe and alpine regions in the north.

Cold desert flora. On the driest uplands of Antarctica and in oases, encrusting grey and orange lichens *Buellia frigida, Alectoria miniuscula, Biatorella antarctica, Umbilicaria decussata* and species of *Caloplaca, Xanthoria, Usnea* and *Candelaria* may be the only visible signs of life. In damper areas *Nostoc commune*, a prominent blue-green alga, occurs widespread in plaques associated with bacteria, green algae, fungi and mosses; in early spring their black, dried-out patches show where summer moisture will gather. Two macrolichen species are shown in Figure 3.10.

Where a slight flow of water is guaranteed for a few days, for example in crevices and gullies that conduct snow-melt trickles, mosses gather among the lichens. Where water accumulates, mosses become dominant; the wettest patches have mats, even hummocks, of *Bryum, Grimmia, Ceratodon* and other species, growing from a substrate of dead but barely decomposed organic material and often partly overgrown with lichens. *Prasiola crispa* and other green algae are widespread at the coast, forming mats where seabird droppings add nutrients to melt-water; again lichens and mosses may grow in association.

Colonization and succession of plants are reviewed in Smith (1984). Both are slow processes; the establishment of a typical antarctic plant community involving only a few species may take centuries or even millennia (Lamb, 1970). In some situations the first colonizers are algae, often nitrogen-fixing

Figure 3.10 Macrolichens (*Umbilicaria decussata*, grey, reticulate, and *Pseudophebe minuscula*, black) on dry windswept boulders and rock faces. Two of the dominant and most widespread plant species on continental Antarctica. Both are bipolar species. Photo: Ron Lewis Smith

cyanobacteria or blue-green algae (*Nostoc*, *Oscillatoria*), followed by lichens and mosses which draw nourishment from the algal detritus (Llano, 1965). In others lichens colonize and replace each other (Lindsay, 1978), in a process taking 200 years or considerably longer.

Oases. Shumskiy (1957) defines antarctic oases as substantial ice-free areas separated from an ice-sheet by an ablation zone, and kept free from snow by ablation due to low albedo and radiation. The most likely cause is a local

Table 3.6 Antarctic oasis areas: after Pickard (1986a).

Oasis	Position	Area (km^2)	Highest point (m)	Topography	References
Dry Valleys	77°S 161°E	>1000	>2000 m	Mountains	Wrenn and Webb (1982)
Bunger Hills (Oasis)	66°S 100°E	482	172	Low hills	Wisniewski (1983), Ròzycki (1961)
Vestfold Hills	69°S 78°E	411	157	Low hills	Pickard (1986)
Greason (Windmill Is.)	66°S 110°E	<40	<100	Low islands	Korotkevich (1971)
Schirmacher	71°S 11°E	23	?	Low hills	Solopov (1969)
Sôya Coast	69°S 40°E	100	500	Peninsulas and islands	Omoto (1977)
Thala Hills (Molodezhnaya)	68°S 46°E	8	<100	Low hills	Aleksandrov and Simonov (1981)

reduction in precipitation. Details of seven antarctic oasis areas of size range 8–1000 km^2 appear in Table 3.6. This list is minimal, for many more ice-free areas around the continent would qualify for inclusion at the lower end of the size range. Oases are misleadingly named; they are simply polar desert areas, notable for their absence of ice but not for floristic abundance. Similar ice-free areas in the arctic go unremarked.

Botanically, oases contain only the alga–lichen–moss communities that would be expected in continental habitats relatively recently freed from ice (i.e. during the last few thousand years), where water is scarce and winds are strong. In the Vestfold Hills oasis on the Ingrid Christensen Coast, East Antarctica, Broady (1986) recorded over 80 species of terrestrial algae, and Seppelt (1986a, 1986b) found 22 species of lichens and four mosses; a further three species of moss and a hepatic were recorded nearby. Botanically this oasis is now one of the most completely studied areas of continental Antarctica (Pickard, 1986b, 1986c).

Maritime Antarctic communities. By Smith's (1984) reckoning (see Table 3.5) Antarctic Peninsula alone has 20% fewer taxa of lichens than continental Antarctica, but 50% more taxa of mosses. The maritime antarctic as a whole has 20% more lichens than the continent and 150% more mosses; in addition it has 25 taxa of hepatics to the continent's one, and two species of angiosperms that have not yet been found on the continent. Thus the maritime zone is richer both in species and in variety of plant communities than the continent, though still poor in comparison with an arctic biome much closer to the pole.

Smith (1984) presents a generalized classification of antarctic vegetation based mainly on maritime communities; a more detailed account of a similar classification appears in Gimingham and Smith (1970). There are two major divisions, a widespread non-vascular cryptogam formation and a herb formation. The former lists seven or eight subformations made up of algae, lichens and mosses in varying combinations; the latter includes only one subformation, involving the two antarctic flowering plants, the grass *Deschampsia antarctica* and the pink or pearlwort *Colobanthus quitensis.*

Almost all the maritime alga–lichen–moss communities are richer in species and more luxuriant in growth than their continental counterparts, more so on the outlying island groups than on Antarctic Peninsula, and more on the northern than on the southern peninsula. Two subformations are particularly prominent. In the *Usnea–Andreaea* association of lichens and mosses the two named genera are usually represented, together with other species of lichens and algae. This is a versatile and widespread subformation common on all kinds of surfaces, in which crustose and foliose lichens combine with moss cushions to form mats (in the most

Figure 3.11 Moss peat bank formed by *Chorisodontium aciphyllum*, reaching 2 m depth, NW Signy Is. Base radiocarbon dated as *c.* 5500y BP (corrected). Photo: Ron Lewis Smith

Figure 3.12 Section through moss cushion (*Bryum algens*) showing annual growth bands (Signy Is.). Photo: Ron Lewis Smith

favourable conditions, carpets) several centimetres deep. The *Tortula–Grimmia* association of mosses occurs on wet alkaline substrates (for example marble outcrops); species of the named genera form deep carpets in which other mosses, lichens and hepatics settle, forming some of Antarctica's most complex stands of vegetation (Gimingham and Smith, 1970). A moss peat bank is shown in Figure 3.11, and a section through a moss cushion in Figure 3.12.

The herb subformation involves one or both species of flowering plants, usually in combination with mosses. It occurs mostly on sheltered, well-watered cliff faces and damp coastal flats where there is local protection from wind and spray. In specially favoured areas the grass is extensive enough to form patches of turf; the pinks are usually rarer and more dispersed. Despite intensive searches neither has been found on Bouvetøya and only the grass is known from the South Sandwich Islands. Elsewhere they are usually found together; the southernmost record is from Barn Island (62°41′S), Terra Firma Islands, Marguerite Bay (Smith and Poncet, 1985).

The periantarctic zone. Antarctic islands that lie north of the northern limit of pack ice have markedly longer summers and milder winters than those that are invested by pack ice. At sea level and on their lower slopes, they have correspondingly richer vegetation with more species (Table 3.4), more complete ground cover, and especially more flowering plants. Warmth, moisture and deep-rooted plants ensures the presence of mature soils on all these island; coastal regions of the three largest are notably green, with rich stands of tussock grasses, shrubs and fellfields. The presence of soils in turn ensures the presence of breeding petrels that burrow in soils, or require the dense vegetation for nesting, and of a few species of birds (and indeed of introduced mammals) that live entirely on the resources of the land.

South Georgia and Heard Island, lying well south of the Antarctic Convergence, are both heavily glaciated. South Georgia, with a wide range of terrestrial habitats below its permanent glaciers, has recruited a flora of considerable variety, mostly downwind from South America. Heard Island, comparatively tiny, has little open ground and a smaller range of habitats; comparative remoteness has presumably given it fewer opportunities to acquire a flora. Iles Kerguelen and Macquarie Island, lying on the Antarctic Convergence, have correspondingly richer floras close to sea level, including several species of ferns and flowering plants.

3.4 Land animal communities

3.4.1 *Land invertebrates*

Arctic land invertebrates. Damp polar desert soils and vegetation mats contain a small but active microfauna of consumers that feed on bacteria, algae, fungal hyphae and plant debris. Soils of Devon Island, for example, yield protozoa, rotifers, tardigrades, turbellarians, nematodes, enchytraeid worms, copepods, ostracods, cladocerans, mites, spiders and insects (Ryan, 1977). In wetter areas platyhelminths, collembolae and larvae of chironomids and other flying insects become dominant. On the warmer tundra beetles, moths, butterflies, ichneumon flies, bumblebees, craneflies, blowflies and other diptera become prominent in summer, with larvae that live in fresh water, moist soil, vegetation, or in the living or dead bodies of other animals. Spiders and mites are widespread; biting simuliid flies and mosquitos plague man and other warm-blooded animals, and warble flies become the special bane of reindeer and caribou.

These small invertebrates form a community of a kind that is familiar in soils and low-standing vegetation of temperate regions at either end of the world. The herbivores are mainly microfeeders, consuming bacteria, algae, fungi and detritus. Conspicuous by their absence are larger invertebrate herbivores, for example browsing and grazing insects, that often become dominant on temperate grasslands. The carnivores are small and relatively insignificant. All are strongly seasonal, with activity limited to a few days or weeks in summer. Polar adaptations of a few species, so far as they are recognized, are discussed in Chapter 6.

Antarctic land invertebrates. The southern assemblies of terrestrial invertebrates that compare most directly with those of the arctic tundra are found, not on continental Antarctica, but on uplands of the larger periantarctic and subantarctic islands. These are the environments that, in terms at least of summer climates, most closely match milder parts of the northern tundra. However, it is not surprising that the southern islands, relatively tiny and lacking contiguity with major landmasses, are comparatively poor in phyla, families, genera and species.

Maritime and continental Antarctica, recently planed by glaciation and with even fewer opportunities for recruitment, are poorer still. Insects, mainly collembola, and arachnids, mainly mites, are the dominant forms (Block, 1984). Maritime Antarctica has in addition one species of soil copepod, two of chironomid midges (Figure 3.13), and a very limited

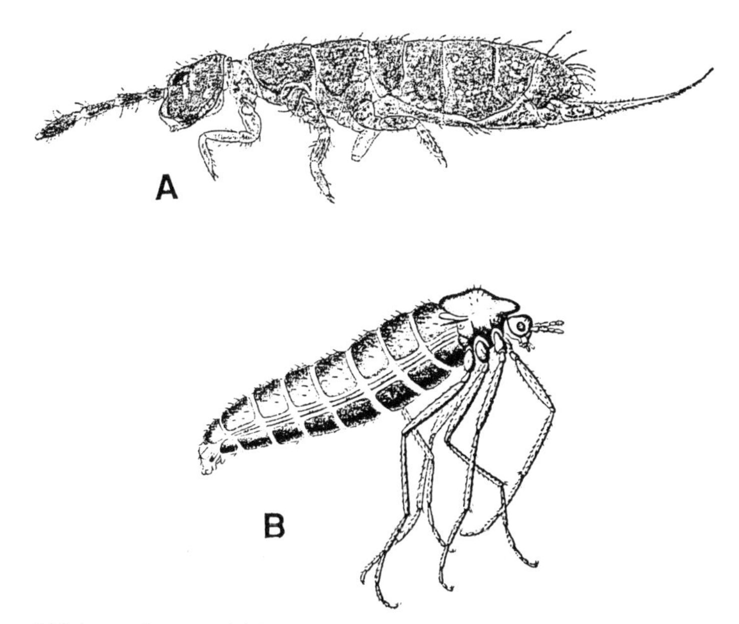

Figure 3.13 Antarctic terrestrial fauna. (*A*) *Isotoma octo-oculata*, a soil collembolan: body length 2 mm; (*B*) *Belgica antarctica*, a coastal wingless midge: body length 3 mm. Collected on Antarctic Peninsula by biologists of the *Belgica* expedition (1897–99).

fauna of introduced species, local to scientific stations and transient. Notable absentees are earthworms and enchytraeids, terrestrial molluscs, amphipods and isopods, hemiptera, lepidoptera, pseudoscorpions, spiders and myriapods (Block 1984). Many of the vertebrates at either end of the world carry resident populations of feather lice, mites and other ectoparasites that breed on the body; nestbreeding fleas and ticks are understandably less prominent in high latitudes, where nest material is minimal.

3.4.2 *Land vertebrates*

Arctic land vertebrates. Amphibia and viviparous reptiles (for example vipers *Vipera berus*) reach the northern edge of the forest but rarely cross the treeline. Birds are prominent on the tundra, especially in summer when migrant waders and waterfowl pour in from the south; only a few species are year-round residents. Mammals are plentiful though less prominent. Voles, mice and lemmings are active under the snow in winter, hunted by predatory weasels, stoats, foxes and wolves. Larger browsers and grazers include hares, rabbits, musk oxen, and reindeer and caribou. Polar adaptations of some of these species are discussed in Chapter 6.

Resident land birds. Arctic land birds breed almost entirely between mid-May and July, when days are long, the weather is propitious, the ground is unfrozen and relatively clear of snow, and food is most easily hunted. Outside the months of breeding all but a few species disappear to wintering grounds well south of the Arctic Circle; thus the arctic has over 150 species of breeding land birds, but only eight of them reside on the tundra all the year round. The resident species include stocks that breed far to the north in summer, and in autumn move south to the warmer tundra, or from tundra to warmer coastlands, to avoid the harshest winter conditions. Three pairs of closely related species—the buntings, redpolls and ptarmigan—each with a northern and southern representative, illustrate graded levels of arctic penetration.

Sparrow-sized buntings and redpolls, weighing 30–35 g, are the smallest and perhaps the hardiest of polar birds. Too small to be able to carry massive winter insulation, they nevertheless survive; it is remarkable to see such tiny objects maintaining normal avian body temperatures (39–40°C) in ambient temperatures some 60–70°C lower, seeking shelter under eaves and in or under snow, and maintaining their high metabolic rate by constant foraging. Lapland buntings *Calcarius lapponicus* winter in temperate and subpolar latitudes and breed on the southern tundra in summer (Tryon and MacLean, 1980). Closely-related snow buntings *Plectrophenax nivalis* winter in small flocks where the snow is thinnest on the southern tundra, and fly northward to breed on the high arctic tundra in summer, when snow still plentiful on the ground (Pattie, 1972, 1977). They maintain themselves throughout the year on a diet of seeds and insects.

Common redpolls *Carduelis* (= *Acanthis*) *flammea* and arctic redpolls *C. hornemanni* make up a similar pair, the former more southerly in distribution but overlapping in parts of their range. Both species feed mainly on birch seed in winter, finding it respectively on snow-free areas of the tundra and in the sub-boreal forests. Breeding also on the antarctic fringe (see below), common redpolls have the widest latitudinal breeding range of any passerine.

Of the third pair, rock ptarmigan *Lagopus mutus* and willow ptarmigan *L. lagopus*, the former breed and winter farther north. Both shift annually from relatively thin brown or speckled summer plumage to much denser white winter plumage (West, 1972). They feed on insects, seed and shoots, digging in the snow with well-feathered feet and burrowing for food and shelter.

Ravens *Corvus corax* and snowy owls *Nyctea scandiaca* (Watson, 1957) are among the hardiest residents; snowy owls have been reported as far north as 82°N on Ellesmere Island in winter (Gessaman, 1978). Both are large enough to carry dense plumage and subcutaneous fat; both are hunters of other birds and small mammals, and ravens also scavenge on the leavings of other species, including man. Snowy owls, northern counterparts of the cosmopolitan eagle owls (*Bubo* sp.) are usually sedentary, wintering on or close to their breeding areas. However, high populations and food shortages, occurring every four or five years (see below), drive them far south into and beyond the boreal forest (Fogden, 1973). They breed from mid-May onward, hatching eight or more young in good years but only one, two or none when food is scarce.

Migrant land birds. Every spring the Arctic's tiny resident populations are swamped by a huge influx of migrant birds, which fly from the tropics and temperate zones to feed and breed on the tundra. Over 120 species are involved, most prominent among them waders (shorebirds), waterfowl including divers (loons), geese, swans and ducks, and birds of prey. Many migrate almost entirely overland, along 'flyways' or traditional routes featuring coastlines, rivers and mountain chains. Their flights may take a few hours or many days.

Some species have long passages over the sea, for example the ducks, geese, waders and passerines that cross from northwestern Europe to Greenland. Lesser goldern plovers *Pluvialis dominica* make long trans-oceanic flights; Alaskan breeding stocks cross 4000 km of the Pacific from wintering in Hawaii, while Asian stocks make even longer journeys from New Zealand and Oceania to breed in Siberia. For a review of migration see Evans (1985); for reviews of breeding patterns and adaptations among arctic migrants see Perrins and Birkhead (1983) and Sage (1986).

Migration is usually initiated as a physiological response to increasing day length. The birds fatten beforehand for their long flight, and unless delayed seriously by bad weather they arrive on the breeding grounds with fat in reserve. This helps to protect them from starvation during the first week or two, when conditions may still be hard. In species such as swans and geese, which lay large clutches of eggs soon after arrival, the reserve on landing may indeed determine breeding success for the year.

The migrants waste no time in starting to breed. Large species with long incubation and growth periods, for example swans and geese, are particularly pressed for time. Many are already paired from previous matings; all are usually nesting, courting, mating and laying within a few days of arrival. Unseasonable bad weather may mean a late start to breeding and the loss of an entire season's young; in particularly bad seasons large numbers of birds may fail even to start, and miss a breeding season altogether. After incubation and the exertions of tending their young, adults require the rest of the summer to regain condition and undergo at least a partial moult, then to fatten again for their journey back to the wintering grounds. In some species the post-nuptial moult occurs in areas separate from the breeding grounds, requiring a midsummer migration.

A tundra bird community. The summer nesting population of Truelove Lowland, on Devon Island, is a typical high tundra assembly; for details of breeding and feeding, including special studies of the snow buntings, see Pattie (1977). Three species of arctic residents, Lapland longspurs, snow buntings and rock ptarmigan, are joined annually by a dozen species of long-distance migrants—red-throated divers (loons) *Gavia stellata*, arctic terns *Sterna paradisaea*, oldsquaw *Clangula hyemalis*, common and king eider *Somateria mollissima* and *S. spectabilis*, glaucous gulls *Larus hyperboreus*, snow geese *Chen caerulescens*, parasitic and long-tailed jaegers *Stercorarius parasiticus* and *S. longicaudus*, white-rumped and Baird's sandpipers *Erolius fusicollis* and *E. bairdii*, and black-bellied plovers *Squatarola squatarola*. The total bird population of the lowland includes occasional-breeding snowy owls *Nyctea scandiaca* and peregrine falcons *Falco peregrinus*.

In general terms, the snow geese and ptarmigan feed mainly on vegetation. Snowy owls and falcons, which nest only occasionally on Truelove Lowland, are the top predators, feeding mainly on small mammals and birds. Between these extremes jaegers, terns, loons, plovers, buntings, longspurs, sandpipers and ducks occupy secondary and tertiary feeding levels, foraging for aquatic invertebrates, insects and seeds. Gulls and jaegers especially are opportunists, feeding on insects, fish, small mammals and whatever else is plentiful.

As in lower latitudes, clutch size and nesting success among arctic land birds depend on the amount of food available each season, and this varies considerably from area to area and season to season. In areas where conditions are favourable, in seasons of good weather, productivity of vegetation and insects may locally be very high. While numbers of residents available for breeding each year depends on local productivity in previous seasons, numbers of migrant birds are determined partly by circumstances and events on their wintering grounds, where food may be scarce and predation high; the birds are also considerably at risk during their long flights. Thus arctic breeding grounds often appear underpopulated in relation to their carrying capacity, and there is plentiful food for all that manage to get there safely. This is what makes the migrants' long journey to the Arctic worth while. In some passerine species it even justifies slightly larger clutches than are found in the same or closely related species of lower latitudes (Ricklefs, 1980).

However, food species are few, food webs are simple, and there are few alternatives if one or two sources of food disappear temporarily. Buffers and balances that in temperate latitudes keep populations steady have less application in polar regions, and swings between success and failure, between underpopulation and overcrowding, tend to be violent. Thus in years when lemmings and other small rodents are plentiful (see below), owls, skuas, jaegers and other aerial predators fare well and raise many offspring. Foxes, weasels and other ground predators also flourish on the rodents, reducing their take of eggs and fledglings, and more of the young birds survive to breed in future years. After two or more good seasons in succession, predatory birds become markedly more plentiful. They nest at higher densities and leave no room for non-breeding birds, which are displaced and forced to 'irrupt' to unfamiliar ground. So snowy owls may appear in southern Canada, Britain and southern Europe (see above), while their populations remaining at home suffer poor breeding seasons due to

Table 3.7 Terrestrial mammals of the Arctic. After Sage (1985). Species marked with asterisk are year-round residents.

	High Arctic	Marginal Arctic	Greenland	Eurasia	Alaska	Canada
INSECTIVORA						
Common shrew *Sorex araneus*				×		
Arctic shrew *S. arcticus*				×	×	×
Laxmann's shrew *S. caecutiens*				×		
Masked shrew *S. cinereus*			×	×	×	
Large-toothed shrew *S. daphaenodon*				×		
Pigmy shrew *S. minutus*				×		
Dusky shrew *S. obscurus*					×	×
Flat-skulled shrew *S. vir*				×		
RODENTIA						
Alaska marmot *Marmota broweri*					×	?
Black-capped marmot *M. camtschatica*				×		
Arctic ground squirrel *Spermophilus undulatus*				×	×	×
*Insular vole *Microtus abreviatus*	×				×	×
Narrow-skulled vole *M. gregalis*				×		
*Middendorff's vole *M. middendorffi*	×			×		
Tundra vole *M. oeconomus*				×	×	×
Meadow vole *M. pennsylvanicus*						×
*Arctic (collared) lemming *Dicrostonyx groenlandicus*	×		×	×	×	×
Hudson Bay lemming *D. hudsonius*	×					×
*Brown (Siberian) lemming *Lemmus sibiricus*	×		×	×		
Grey red-backed vole *Clethrionomys rufocanus*				×		
Northern red-backed vole *C. rutilus*				×	×	×
Eastern vole *Eothenomys lemminus*				×		
European water vole *Arvicola terrestris*		×		×		
Muskrat *Ondatra zibethicus*						×
Porcupine *Erithrizon dorsatum*		×			×	×
LAGOMORPHA						
Snowshoe hare *Lepus americanus*	×				×	×
*Arctic hare *L. arcticus*	×		×			×
*Alaskan hare *L. othus*	×					
*Varying hare *L. timidus*	×					

Table 3.7 (*Contd.*)

	High Arctic	Marginal Arctic	Greenland	Eurasia	Alaska	Canada
Northern pika *Ochotona hyperborea*		×		×		
CARNIVORA						
Coyote *Canis latrans*	×				×	×
*Gray wolf *C. lupus*	×		×	×	×	×
Red fox *Vulpes vulpes*				×	×	×
*Arctic fox *Alopex lagopus*		×	×	×	×	×
Grizzly bear *Ursus arctos*				×	×	×
*Polar bear *U. maritimus*	×		×	×	×	×
*Stoat (ermine) *Mustela erminea*	×		×	×	×	×
*Least weasel *M. nivalis*	×			×	×	×
European mink *M. lutreola*		×		×		
American mink *M. vison*		×			×	×
*Wolverine *Gulo gulo*	×		×	×	×	×
Otter *Lura canadensis*		×			×	×
Lynx *Lynx canadensis*		×			×	×
ARTIODACTYLA						
Moose/elk *Alces alces*				×	×	×
*Caribou/reindeer *Rangifer tarandus*	×		×	×	×	×
*Muskox *Ovibos moschatus*	×		×	×	×	×
Dall sheep *Ovis dalli*					×	×
Snow sheep *O. nivicola*			×			

overcrowding. The subsequent collapse of the rodent population due to competition and over-predation leads in turn to the collapse of aerial predator populations, temporarily restoring an unstable balance.

For more general acounts of Arctic land birds see Stonehouse (1971) and Sage (1986).

Arctic land mammals. Of some 4000 species of mammals in the world as a whole, only about 50 species have penetrated the Arctic (Sage, 1985 and Table 3.7). This list includes eight marginal species—mountain and forest mammals that occur only on the arctic fringe—and over a dozen others that seldom penetrate far and are never found in the northern zones. Just over a dozen species, marked with as asterisk in Table 3.7, are widespread on the tundra throughout the year.

These are mostly species of wide arctic distribution, morphologically distinct from their nearest kin outside the Arctic, suggesting a long-term

Figure 3.14 Dall sheep, an arctic mountain species confined to relatively snow-free areas. Photo: B. Stonehouse.

commitment to polar life. All are lowland species whose ancestors probably inhabited lowland forests and steppe; curiously absent from the polar tundra are montane mammals of the subpolar fringe, for example Dall (Figure 3.14) and snow sheep, marmots, pikas, and upland (Siberian and red-backed) voles, which might have been expected to take readily to life on the low tundra (Chernov, 1985).

Largest among the tundra herbivores are musk oxen and caribou. Too big to seek shelter within the snow, they are equipped with unusually dense fur and live in groups which offer protection equally from cold and predation. Musk oxen were formerly widespread across the low tundra, but hunting by man has now restricted them mainly to north Greenland and the Canadian far north. They feed voraciously in summer and store energy as subdermal fat. In winter they keep to high, snow-free ground where continued grazing is possible, losing weight until the new flush of growth in spring. Caribou and reindeer, respectively new-world and old-world representatives of a single species, form large migratory herds which winter on southern tundra or within the forest, and move north across the tundra in summer. Almost all reindeer are now domesticated. Generally smaller and more manageable than caribou, they are widespread in managed herds in southern Greenland and across northern Eurasia. Small herds introduced into northern Canada and Alaska during the twentieth century are now flourishing and helping local economies. Caribou herds, formerly subject to severe hunting, continue to wander across the North American tundra subject to much lighter hunting pressures and non-intrusive human management.

Among large predatory mammals, polar bears are mainly maritime, spending much of their time along the coast or on sea ice where they hunt for fish and seals. Grizzly and Kodiak bears (brown bears) are land-based and mainly herbivorous, feeding on roots, shoots, berries, wild honey and occasionally taking fish, small mammals and birds, and offal. Grey or timber wolves are major predators of small mammals and birds, and also of caribou, reindeer and other deer (for example moose) that emerge onto the tundra in summer.

Small mammals avoid the harshest winter conditions by living within or under the snow. Insectivorous shrews are confined mainly to the forest-tundra and tundra edge; only the arctic shrew and masked shrew, both holarctic species, are widespread on the southern tundra, and none penetrates far to the north (Bee and Hall, 1956). Voles and lemmings, which are mainly herbivorous, range widely across the central and southern tundra, but only three species—arctic, Hudson Bay and brown lemmings—are plentiful on the northern tundra. Of the lagomorphs, snowshoe hares are animals of forest and forest-tundra; Alaskan, varying and arctic hares are widely-ranging tundra species (some taxonomists say merely races of a single species). Some southern populations winter in the forest; northern populations winter in the snow but are often found further north than any other mammals (Anderson and Lent, 1977). Predatory on the smaller species are wolverine, arctic foxes, otters, mink, weasels and short-tailed weasels (ermine), which take small mammals, birds, eggs and carrion.

Wintering within the snow's protection (see summary in Sage, 1986b), feeding on last year's vegetation at the snow–ground interface, lemmings and other small herbivores with high capacity for reproduction are sometimes able to produce four or five litters per year, each of five or six offspring. After two or three such years their numbers build up spectacularly within favoured areas. Though owls, skuas and other predators move in and often enjoy successful breeding at their expense, the proliferating herbivores continue to outnumber them and to outpace the relatively low productivity of their food supply.

When they have eaten much of the available food, the huge population collapses; though some individuals emigrate, most die of stress, diseases and starvation. Because they are likely to have damaged storage organs and growing points as well, the capacity of the plants to recover is slowed. Thus population numbers of small herbivores, in circumstances of limited food production, cycle over periods of three to four years. Those of some larger mammals, for example hares and lynx, fluctuate in longer cycles of 10–13 years. This phenomenon, characteristic of arctic populations but not confined to them, is discussed by Batzli *et al.* (1981) and Crawley (1983).

Antarctic land birds. There are no resident land birds in maritime or continental Antarctica. Vagrant birds of South American origin are reported from time to time on the South Orkney and South Shetland Islands; their presence indicates repeated possibilities for colonization, but they invariably disappear quickly. Islands within the northern limit of pack ice support only two species on land—sheathbills *Chionis alba* and *C. minor*. Though land-based and pigeon-like in appearance, sheathbills have maritime attributes of dense plumage, including thick grey underdown, and a store of subdermal fat. They stay close to the sea, scavenging on penguin, cormorant and seal colonies, along tide lines, and in rock pools (Jones, 1963; Stonehouse, 1985). In the periantarctic zone some individual *C. alba* are year-round residents on South Georgia; others migrate, presumably to the South American mainland. The small population of *C. minor* on Heard Island is entirely resident, not surprisingly so in view of the distance over water to more agreeable wintering grounds.

South Georgia supports also resident pipits *Anthus antarctica* and pintails *Anas georgica* (Weller, 1975), both derived from similar South American stocks. These feed on insects, shore-line and tide-pool fauna and soft vegetation in South Georgia's deep sheltered fjords. Breeding speckled teal *A. flavirostris*, also of South American origin, were first recorded on South Georgia in 1971 (Weller and Howard, 1972). Macquarie Island, close to the Antarctic Convergence, has a resident population of common redpolls *Acanthis flammea*, a species which spread south to subantarctic islands after its nineteenth-century introduction into New Zealand from Europe. Macquarie Island has also resident grey duck *Anas superciliosa*, mallard *A. platyrhynchos* and weka *Gallirallus australis*, the latter a flightless rail introduced from New Zealand in the late nineteenth century. Iles Kerguelen, also close to the Convergence, support an endemic pintail *A. eatoni*, closely akin to the northern pintail *A. acuta* (Watson *et al.*, 1975).

Antarctic land mammals. Though Antarctica must once have been the home of marsupials and possibly of placental mammals, it is currently the only world continent without a land mammal population. None of the peripheral islands has naturally-occurring land mammals, but many have acquired rats and mice accidentally and browsing and grazing mammals by design; Leader-Williams (1985) lists 15 species introduced to the islands. Rodents came ashore mostly from sealing and whaling ships; they survive well, mainly close to sea level, on all the main island groups north of the northern limit of pack ice. On South Georgia reindeer, introduced from Scandinavia by whalers early in the twentieth century, flourish and appear to be spreading; horses and sheep introduced to the same island did not survive. Farming and stock-raising were attempted on Auckland and Campbell Islands (Fraser, 1986) and Iles Kerguelen (Lesel and Derenne, 1975), but were never successful for long.

3.5 Summary and conclusions

Cold polar land environments are relatively new; though cold mountain environments may have existed throughout the earth's history, until the current glacial period polar regions have been temperate at sea level. Climatic fluctuations have ensured that most polar soils are recently formed; because of ice-induced turbidity and slowness of soil-forming processes in cold conditions, the soils remain mostly immature, and unpromising media for the growth of vegetation. Arctic soils show the wider spectrum of maturity; even in the most favoured conditions, few antarctic soils have passed beyond the ahumic stage. Arctic terrestrial vegetation, drawn widely from the north temperate region, varies widely in quality, density and productivity from south to north. Antarctic vegetation is comparatively meagre, including for example only two species of vascular plants. Similarly, comparisons between northern and southern populations of land animals reflect the many more opportunities for polar recruitment, and the wide range of available niches, to be found in the Arctic.

CHAPTER FOUR
FRESHWATER ENVIRONMENTS

4.1 Fresh water in polar regions

The chronic aridity of both polar heartlands has already been noted. Though the antarctic ice-cap is estimated to contain 90% of the world's water, only a tiny fraction of it becomes available each year for living material. Similarly the arctic desert, with or without ice sheets, has very little water available during the short summers, and none in winters. On the polar fringes warmer air brings in moisture and deposits it as hoar-frost, snow or rain. Though water is solid for much of the year, the annual thaw in spring and summer provides a wide range of aquatic habitats that support biological activity, and may even allow high productivity for a few weeks each year.

Wet soil is essentially a freshwater habitat, dealt with in Chapter 3 above. Melting snow and ice surfaces provide transient substrates which bacteria and algae colonize, forming patches of pale green, red or yellow several hectares in extent. Snow-melt and rain gather in runnels, streams and even small rivers, that flow for a few hours, days or weeks in summer but are stilled by freezing for the rest of the year. Ponds and lakes form in poorly-drained rolling country, accumulating debris and nutrients, freezing partially or completely in winter but providing a limited habitat for plants and animals in summer.

4.2 Polar freshwater habitats

4.2.1 *Snow and ice habitats*

Freshly-fallen snow surfaces are virtually sterile, though bacteria are present in detectable quantities even on the southern polar ice-cap (Meyer-Rochow, 1979). Older snow surfaces become adulterated; damp snow attracts wind-blown particles electrostatically (Benninghoff and Benninghoff, 1985), acquiring films of rock dust, bacteria, spores and plant propagules. These accumulate as the snow melts under spring sunshine, and the resulting discoloration hastens the melting. Fragments of moss and algae that land on snow surfaces survive well in the moist environment, and may settle through the remaining snow to colonize the ground beneath.

Ionization and other atmospheric processes add nitrates and ammonium compounds to the snow in nutritionally significant quantities (Parker *et al.*, 1978). Snow close to the coast is often additionally contaminated by mineral salts and organic material from the sea. In summer these are carried inland in sea spray; in winter impregnated snow from the sea ice deposits them on coastal rocks and snowdrifts. In late spring, snow surfaces within a few hundred metres of the coast often show a flush of colour, usually pink, green or brownish-yellow, caused by patches of unicellular or colonial algae. On the South Orkney Islands *Chlamydomonas, Raphidonema* and possibly *Ochromonas* are implicated (Fogg, 1967), similar accumulations of algae in a wide range of species have been seen at many other points around the coast (Akiyama, 1979; Kol and Flint, 1968; for a recent review see Vincent, 1988). Their productivity is low, fixed carbon being of the order of 10 mg m^{-2} of snow per day; colour differences are due to different stages in development of the cells, and their sudden, often dramatic appearance during warm spells is more probably due to progressive loss of covering snow than to rapid proliferation (Fogg, 1967). Snow algae are widespread in alpine and polar regions, but require persistent banks of melting snow to form spectacular coloured patches; they are rarely seen, for example, on the colder coasts or interior ice-cap of Antarctica. Their ecology has been reviewed by Hoham (1975, 1980).

On the lower reaches of glaciers and on ice shelves close to the sea, where ablation exceeds snowfall in summer, wind-blown dust and rock fragments encourage local melting. The resulting shallow pits and pools, often extending over many hectares, accumulate salts and nutrients from year to year. Many are rich enough to support a distinctive cryoconite flora of algae and cyanobacteria, which survive winters encapsulated in the ice (Wharton *et al.* 1981; Vincent, 1988) and flourish briefly during the summer thaw.

4.2.2 *Still-water habitats*

Over much of the Arctic, surface drainage is poor; permafrost seals the subsoil, and snow-melt provides a flush of water in late spring. In the northern desert there is rarely enough snow for melt-water to accumulate, and ponds are rare. Where winter snow is plentiful, ponds and shallow lakes occupy large areas; over much of the tundra they form networks of static waterways that make walking difficult after the thaw. On ice-scoured uplands and coastal flats there may be more ponds and lakes than dry land. Coastal plains often feature long, shallow lakes, separated and aligned by raised beaches and occupying up to 90% of the terrain. Antarctica has relatively little standing water. Large lakes, many of them strongly saline, are reported from several of the oasis areas; freshwater lakes are also present even in high latitudes, and temporary lakes form in summer alongside many glaciers. Small coastal summer ponds are rare on the mainland but plentiful in the maritime and peripheral Antarctic regions.

Lakes form also on poorly-drained ice, alongside glaciers and ice-caps, on snow-covered ice-sheets, and between ice-sheets and ice-filled moraines. Some are long-lived, fed by annual influxes of melt-water. Others are temporary, flooding during the summer melt and liable to sudden drainage through rifts in the ice and other causes (Hambrey, 1984).

Polar lakes and ponds are partly or completely frozen and snow-covered for much of the year. Large ponds and lakes, especially those in high latitudes, stay frozen all the year round, accumulating snow-melt water from rocks nearby and part-thawing only in exceptional seasons. Large lakes thaw round the edges but tend to retain a core of ice in summer; in consequence their waters never warm far above freezing point. Small lakes and ponds thaw completely. Lakes that owe their volume to a balance between inflowing and outflowing water may change radically between seasons, almost draining in winter when the inflow slows down, and overflowing when summer melt brings in water that cannot escape through still-frozen outlets.

Lakes that are ice-free in summer begin cooling at their surface late in the season. The cooled waters sink and circulate until the water mass reaches the critical temperature of 3.98°C, at which fresh water is densest. Thereafter cooler waters overlie warmer, and the surface freezes while bottom waters are still relatively warm. Shallow ponds freeze to the bottom in winter; deeper ones may remain unfrozen at depth, providing a still, protected habitat for overwintering fish, larvae, eggs and spores.

Conversely, spring warming begins immediately under the ice as soon as sunlight penetrates, usually, in early summer after the surface snow (which is relatively opaque to solar radiation) has ablated or melted. Water that is warmed a degree or two above freezing point sinks into the colder water below, and is replaced; the resulting density currents stir the lake into turbulence and spread warmth, a process that is enhanced by wind-stirring as the ice disperses. Thus the water quickly becomes isothermal, at temperatures close to those of the air or ground surface nearby. Some stratified hypersaline lakes are permanently warmer in their lower layers than at their surface, sometimes because of volcanic heating, more often due to solar warming through the ice and upper layers of water.

Most polar lakes are highly oligotrophic, with clear waters and a very limited range of species inhabiting them. Notable exceptions are arctic oxbow and other remnant lakes formed on river plains, which may be muddy and well endowed with minerals. Ponds range from oligotrophic to highly eutrophic, according to their situation and history. The poorest are the clear waters of ice-scoured uplands, where there is little soil or vegetation. Those on low-lying land invested by tundra vegetation are usually richer, with a seasonal turnover of minerals and slightly higher productivity. However, such important nutrients as nitrates and phosphates are generally scarce, because rates of bacterial decomposition in the soils of the catchment areas are low, and very little leaches out into the water.

Ponds and lakes that are frequented by birds often have an enhanced productivity, based on nutrients from the birds droppings and decaying feathers. Even the richest polar ponds and lakes have fewer species and lower productivity than most ponds of similar size and shape in temperate latitudes.

A few polar lakes are saline, their salts originating either from sea water and spray or from rock weathering in their own catchment areas. Sea water on chilling precipitates salts sequentially, first calcium carbonate, then sodium sulphate at $-8.2°C$, sodium chloride at $-22.9°C$, and magnesium and potassium chlorides at $-36°C$, leaving a residual solution rich in calcium chloride (Nelson and Thompson, 1954; Thompson and Nelson, 1956). Deposits of many of these minerals, notably halites (hydrated sodium chloride) and mirabilite (hydrated sodium sulphate) occur in the sediments and surrounds of many existing lakes, and in dried-out lake deposits. Saline water reaches its maximum density at temperatures below 3.98°C; at 27 ppt and above the maximum is reached at freezing point, which is also determined by salinity.

4.2.3 *Running-water habitats*

Few rivers start in polar regions, though some very large ones flow across arctic and subarctic lands to discharge into the Arctic Ocean. The great northward-flowing rivers of Siberia and the lesser ones of North America (Figure 4.1) bring year-round flows of water and warmth from the south. They bear thick ice in winter and their flow is strongly seasonal (Table 4.1), especially so in those that rise or have much of their catchment area in tundra or boreal forest. Over 35% of the annual flow of the Lena, in

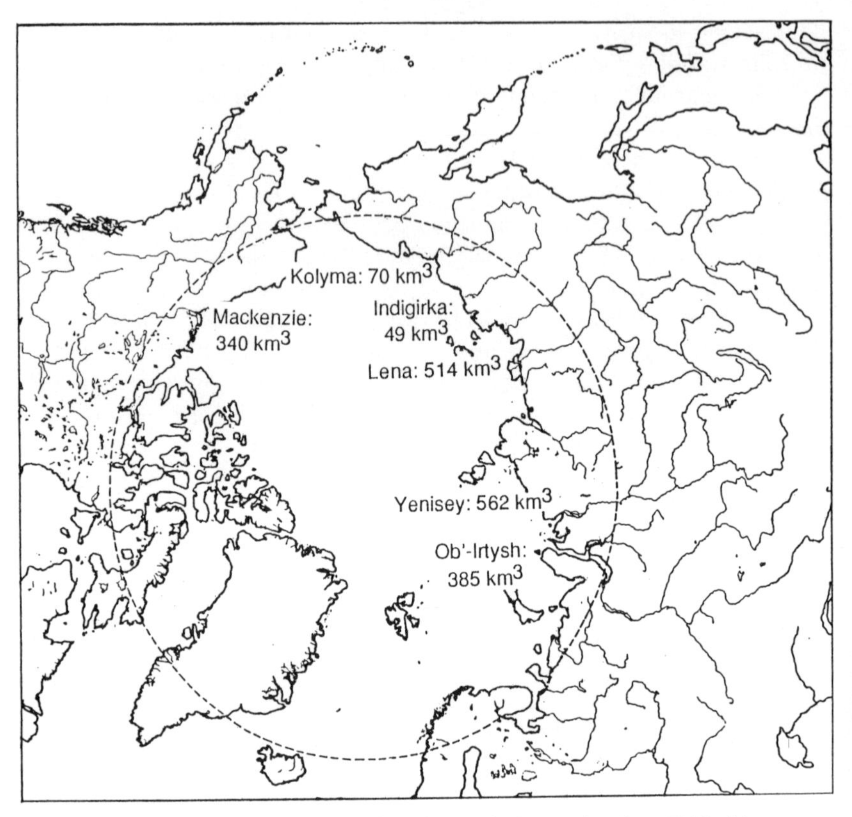

Figure 4.1 Major inputs of fresh water into the Arctic Ocean: data from Table 4.1.

Siberia, is delivered during the peak month of June, less than 1% in each of the three months of February, March and April.

More characteristic of the regions are small graded rivers and streams fed by springs and surface runoff, and seasonal, short-lived torrents that flow from glaciers. In these up to half the annual output may occur in the three or four weeks of springtime thaw. Like the larger rivers they are erosive, undercutting and destroying their banks to form wide, braided channels. They carry heavy loads of sediment, abandoning it along their course and in extensive deltas; the finer deposits are often colonized by vegetation, providing some of the richest and most productive communities in the tundra mosaics. More characteristic still are the thousands of tiny melt-water streams that trickle from ice-sheets and snow-banks for a few days or

Table 4.1 Annual and quarterly flows of six major river systems into the Arctic Ocean. After Mackay and Löken (1974)

	Jun–Aug	Sep–Nov	Dec–Feb	Mar–May	Annual flow (km^3)
Ob'-Irtysh	57	20	9	14	385
Yenisey	57	18	7	18	562
Lena	71	22	4	3	514
Indigirka	85	14	< 1	1	49
Kolyma	73	17	2	8	70
Mackenzie	38	27	19	16	340

weeks each summer, often following the same channels year after year, and sometimes building up mats of vegetation along their course.

The Arctic is well endowed with running-water habitats. Continental and maritime Antarctica have no large year-round rivers, and only a very few smaller ones, mostly glacier-fed torrents, to match the smaller tundra rivers of the north. The periantarctic islands have many permanent or seasonal streams, spring-fed and carrying run-off from the peaty vegetation throughout the year.

4.3 Aquatic ecology: food webs and productivity

4.3.1 *Arctic aquatic communities*

Still-water communities. Arctic ponds and lakes freeze to depths of 1–2.5 m from September onward, and accumulate a layer of compacted snow on top of the ice during winter. Thawing occurs slowly in May and June; generally the larger the water body, the longer it takes to thaw out in spring, and the cooler it remains through summer. In a study of lakes and ponds on Truelove Lowland, Devon Island (75°30′N), Minns (1977) found the smallest lake (16 ha) free of ice by 17 July, while a larger one of 95 ha retained ice until early August. Temperatures at the lake-beds in about 6 m depths were higher than at the surface when ice was present, rising from 1°C in mid-May to about 3°C in late July. When most of the ice had gone, wind stirred the waters; by late summer temperatures became a uniform 4–5°C at all depths. Temperatures of nearby ponds rose higher, to 8°C.

Char Lake (74°42′N), the subject of long-term studies on Cornwallis Island, with an area of 52.6 ha, was typically ice-free from early August to mid-September. The waters were layered from October to May while ice was present, with temperatures close to freezing point near the surface but at 1–1.25°C near the bottom in 26 m. From June onward warming through the ice set up circulation currents which gradually raised temperatures to a uniform 4°C throughout the water column (Rigler, 1974).

Biological activity starts long before the ice has melted. In Truelove Lowland lakes aquatic algae and bacteria begin photosynthesis as early as February, when snow still mantles the ice and less than 1% of available light penetrates (Hobbie, 1962, 1964; Holmgren, 1968). Often the algae accumulate in a thin layer on the underside of the ice sheet, dispersing when the ice melts. Measuring chlorophyll *a* in water samples as an index of phytoplankton abundance, Minns (1977) found it already plentiful under the ice by mid-May and reaching peak values in late May and June; Kalff *et al.* (1972), working on Char Lake and a neighbouring lake in the Canadian Arctic, recorded active production from February onward.

Algae that are capable of photosynthesis at very low levels of light intensity proliferate early in the season, to be replaced by a succession of other species as light intensity increases (Kalff, 1970). Primary productivity is generally low, averaging 10–50 mg C m^{-2} daily in May and June and rising to 70–80 mg C m^{-2} daily during the summer peak of production; on average only 6–12 g C m^{-2} fixed throughout the year (Kalff, 1970; Minns, 1977).

Lack of nutrients, rather than low temperature or incident radiation, is the single most important factor restricting primary production. Production is often highest in early summer when nutrient levels are maximal, though the water is still close to freezing point. By the time the water warms, all available nutrients have been absorbed and productivity falls. Only a few lakes show a second, late-season peak of production. Shallow lakes sometimes have a bottom layer of aquatic mosses, which by photosynthesis add oxygen to the water close by. Deeper lakes are likely to have the bottom sediment-covered, and the lower water layers may remain unstirred and stagnant.

The fauna of these lakes is usually restricted to rotifers, tardigrades and other small detritus-feeding invertebrates, a few species of copepods and other crustaceans, enchytraeid worms and the larvae of chironomid flies. The planktonic copepods *Limnocalanus macrurus* produce eggs from September to November; the nauplii which hatch after about one month mature in the following June (Roff, 1972, 1973; Roff and Carter, 1972). Mysids *Mysis relicta* by contrast have a two-yearly cycle; females require the same amount of energy for breeding as individuals of the same species in warmer climates, but take a year longer to accumulate it (Lasenby and

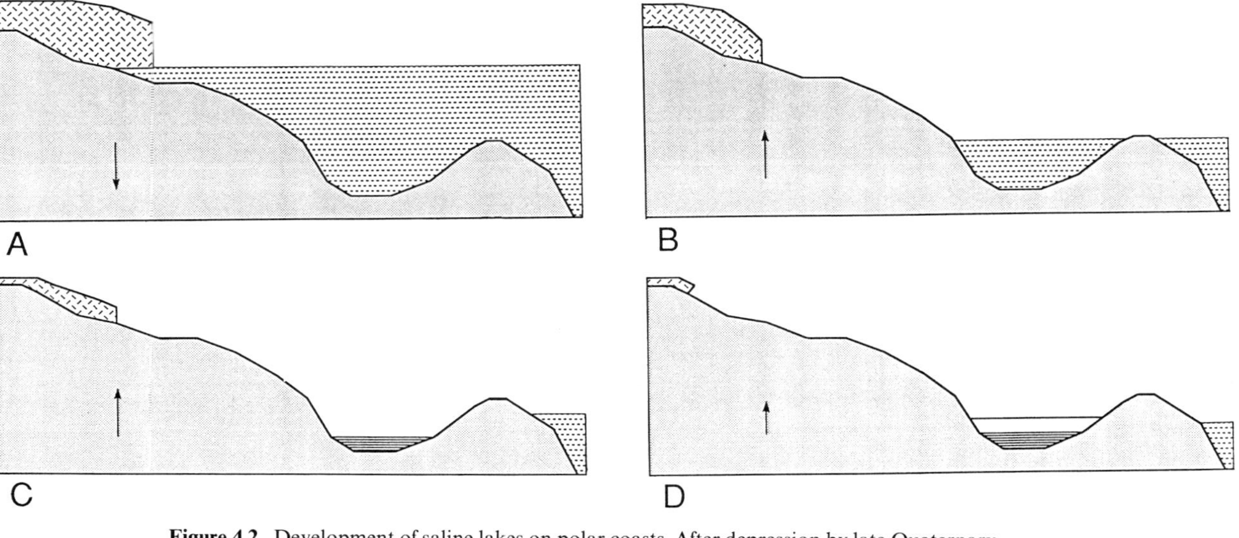

Figure 4.2. Development of saline lakes on polar coasts. After depression by late Quaternary ice caps (A), the land rises, isolating hollows and arms of the sea (B). In dry areas, salinity of the isolated water rises through evaporation and other causes (C), producing (D) a dense bottom layer that subsequent inflows of fresh water fail to disturb. After Dickman and Ouellet (1987).

Langford, 1972, 1973). Chironomids, including several species which bite man and other warm-blooded animals, are plentiful in the lake, forming the main food of arctic char, the single species of fish. Faunal relationships in Char Lake are reviewed by Rigler (1974): studies of individual species or groups include those of Morgan and Kalff (1972) on bacteria, Nurminen (1973) on oligochaetes, Welch (1973) on chironomid larvae and Holeton (1973) on arctic char.

Hypersaline lakes are fairly common in coastal areas of the Arctic which have until recently been ice covered, and since liberation have risen isostatically from below sea level (Figure 4.2); see Hattersley-Smith and others (1970). On Garrow Lake, a hypersaline lake of 418 ha on Little Cornwallis Island, northern Canada (75°N), Dickman and Ouelett (1987) found an ice cover up to 2.4 m thick lasting almost 11 months of the year. With a maximum depth of 50 m and little opportunity for wind-stirring, the lake waters are strongly layered. A brackish upper layer of about 10 m is underlain by about the same depth of more saline water; beneath from about 20 m lies a permanently stagnant bottom layer of ancient sea water—a memento of the island's emergence from the sea some 3000 years ago (Figure 4.3).

During the brief period of open water in summer temperatures rise 2–3°C in the upper 10 m layer. The unstirred brackish water beneath is permanently warmer, temperatures rising to 9.1°C at 20 m, the junction with the deepest layer. From this level downward the temperature falls to 6.5°–7° at the bottom. The anomalous heat is thought to be due to solar energy, which penetrates the surface ice in spring and warms the depths. Phytoplankton and zooplankton occur mainly in the upper 10 m. Diatoms, notably *Cyclotella* spp., predominate in summer, and microflagellates *Rhodomonas* and *Cryptomonas* spp. for the rest of the year. A layer of photosynthetic bacteria lives permanently on the boundary between brackish and highly saline water. Some 51 taxa of algae have been recorded from Garrow Lake, but only one crustacean species *Limnocalanus macrura*, and a single fish species, the four-horned sculpin *Hyoxocephalus quadricornis*.

Running-water communities. Craig and McCart (1975) divide northern Alaskan streams and small rivers into three categories; mountain streams, spring streams and tundra streams. Mountain streams are usually the longest, originating from springs on high ground and picking up ground water and tributaries as they descend. Most flow for four or five months of the year from May to October; some flow throughout the year. Their upper waters remain cool, seldom above 4–5°C in summer; lower reaches may

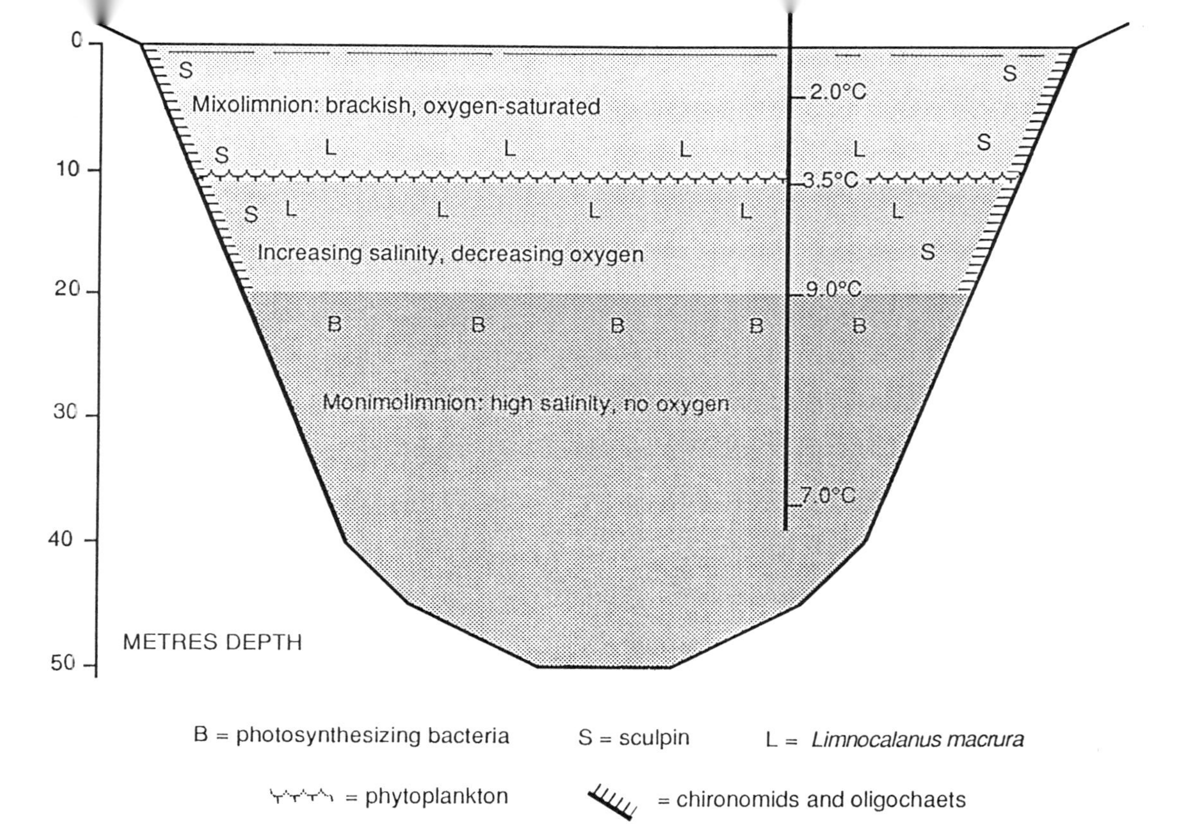

Figure 4.3 Schematic diagram of life and environmental conditions in Garrow Lake, a meromictic saline lake of the high Arctic. After Dickman and Ouellet (1987).

be warmer in summer after crossing the plains, achieving 10–15°C.

Spring streams are generally shorter, typically forming the tributaries of mountain streams. Arising from perennial springs, they are usually small, well established and stable, with high mineral content; many flow year-round. The spring water temperatures remain low and fairly constant throughout the year, seldom falling to freezing point in winter or rising above 8–11° in summer. Tundra streams flow in meandering channels through tundra vegetation, draining the peaty soils, picking up humic acids and acquiring a low pH as they go. Frozen solid in winter, they warm quickly in spring sunshine and may reach temperatures of 15–16°C by high summer, but are the first to slow down and freeze in autumn.

Of the three kinds, mountain streams are least productive, with the smallest standing crop of invertebrates and least variety of species. Tundra streams are richer in invertebrates by a factor of 10, spring streams richer still by a factor of 100. Mountain and spring streams support populations of arctic char *Salvelinus alpinus*; tundra streams support grayling *Thymallus arcticus*, which spawn in their sluggish waters.

4.3.2 *Antarctic aquatic communities*

Still-water communities. Antarctic freshwater ecosystems include permanently frozen lakes high among the inland mountains, seasonal inland or near-coastal lakes of the ice-free areas, some of which are highly saline, and fresh or partly-saline coastal lakes and ponds (Heywood, 1977, 1984). Priddle and Heywood (1980) regard antarctic lakes as forming an evolutionary series which has developed over the period of about 20 000 years since the ice-sheet was maximal; this is in any case a convenient way of describing them.

Proglacial lakes. The simplest lakes lie on or in front of ice-sheets, formed by seasonal melting that may be enhanced by the presence of scree or wind-blown dust. In Antarctica many such lakes occur inland, some on shelf ice close to sea level, others high among mountains or nunataks. Of hundreds recorded from the air, only a very few have been visited even briefly by sledging parties, and little is known of their biology. They accumulate from annual melt-water off the mountains and ice-sheets; their temperatures remain close to freezing point during a superficial summer thaw, though some of the deeper ones may have a permanently unfrozen core. Nutrients are extremely limited, and the lakes support only thin populations of algae and bacteria.

Lakes formed in rock basins by retreating ice-sheets pass through several evolutionary phases. As a moraine-dammed valley is exposed, water flowing in seasonal torrents accumulates on the valley floor, forming a lake that lengthens and broadens as the ice face retreats. The presence of a lake slows the inflow of water, causing rock debris and finely-ground rock flour from the base of the melting ice-sheet to be deposited as sediment on the lake-bed. The surrounding catchment area at first contributes little but clear run-off, and the lake water remains oligotrophic for a long period. However, as vegetation and mineralized soils develop in the catchment, minerals, organic nutrients and soil microorganisms begin to enter the lake from this source. Lakes close to the sea acquire minerals faster than those inland, from wind-blown spray and from seals, penguins, gulls and skuas, which like to wash in fresh water.

Adamson and Pickard (1986) describe several kinds of freshwater lakes in the Vestfold Hills oasis area. Pelite and Chelnock lakes, respectively supraglacial and proglacial, are fed by melting ice cliffs; Crooked Lake is an open-valley lake fed by a through-flowing river, and there are many closed-valley lakes reflecting the water balance prevailing over previous centuries; one, Thalatine Lake, has over 40 minor beach terraces reflecting some 1300 years of evaporation. Others, for example Watts Lake, began as marine inlets but were isolated by isostatic uplift and washed free of salt water by melting land ice; there are also saline lakes of similar origin (see below). Many other antarctic lakes are known at these early stages of development. Few on mainland Antarctica gain substantial nutrients from the soils of their catchment, except around the coast where birds and seals become important agents of eutrophication (Goldman, 1970).

More advanced lakes are found in the maritime Antarctic. Long-term studies of 17 lakes on Signy Island (Heywood *et al.*, 1980; see also Hawes, 1985) provide the most thorough coverage of a wide variety of lake types. Heywood (1984) summarizes the ecology of three lakes of varying nutrient content, occupying valleys of a coastal lowland. These typify many other lakes of similar size and aspect throughout the maritime Antarctic (Heywood, 1987). For studies of similar lakes on the South Shetland Islands see Paggi (1987) and Kieffer and Copes (1987).

Formed in ice-cut cirques or shallow, steep-sided valleys, the Signy Island lakes are moraine-dammed and often rimmed by a rubble-covered shelf about 1 m below the surface; the shelves are swept clear of sediment by wind-induced turbulence, and fall away to a narrow, sediment-filled trough. There is no emergent vegetation. In an environment of heavy snowfall, where mean monthly temperatures are above freezing point in summer but

Figure 4.4 Emerald and Twisted Lakes, Cummings Cove, SW Signy Is. Photo: E. Lemon, BAS

down to $-15°C$ in winter, the lakes are frozen over with 1–2 m of snow-covered ice for 8–11 months each year. Temperatures at the bottom, down to 15 m in the deepest, fall to about 1°C in winter. In summer when melting is well advanced, persistent winds stir the waters and the lakes become isothermal, with temperatures up to 6°C. During this period they receive varying quantities of nutrients and bacteria (Ellis-Evans and Wynne-Williams, 1985) and yeasts (Ellis-Evans, 1985a and b), washed in from their catchments (Table 4.2).

Oligotrophic lakes. Moss Lake is typical of oligotrophic, high-level lakes, recently formed and remote from the sea. In a glacial cirque basin 48 m above sea level and 800 m from the shore, it has a surface area of 15 000 m^2 and a mean depth of 3.4 m. Most water and ions enter the lake directly from precipitation or snow-melt. There is little leaching from the catchment area, which is small and covered with perennial ice, bare rock or scree, with hardly any mineral soil or vegetation. The waters are clear, and salinity and nutrient levels are very low (Table 4.2).

Phytoplankton of diatoms and other green algae is poorly developed; chlorophyll *a* values of up to 8.0 $\mu g\ l^{-1}$ were recorded, and annual

Table 4.2 Salinity and nutrients of three Signy Island lakes: S = summer, W = winter. Data from Heywood (1984).

	Moss S	Lake W	Heywood S	Lake W	Amos S	Lake W
Chlorinity ($mg\,l^{-1}$)	23.0	49.0	35.0	56.0	37.1	94.7
Nitrite N ($\mu g\,l^{-1}$)	1.0	4.1	5.3	21.0	204.0	85.0
Nitrate N ($\mu g\,l^{-1}$)	21.0	50.0	39.0	128.0	321.0	721.0
Orthophosphate P ($\mu g\,l^{-1}$)	0.5	0.4	15.9	58.0	597.0	260.8
Silicate Si ($\mu g\,l^{-1}$)	143.3	267.0	174.0	368.0	191.6	656.9

production was estimated at 12.6 g C m^{-2} in a year (Ellis-Evans, 1981a). In compensation, these clear lakes often have a rich benthic flora of perennial mosses (*Calliergon, Drepanocladus* spp.) and algae (for example *Tolypothrix, Plectonema* and *Phormidium*), both liberally covered with epiphytic blue-green and green algae. Growing slowly to form dense mats, some with strands several m long, these fix up to 6 g C m^{-2} annually. The calanoid copepod *Pseudoboekella poppei* is a planktonic browser, fed on in all stages of its development by the larger copepod *Parabroteas sarsi*. On the lake-bed cladocerans *Macrothrix hirsuticorna*, *Aloma rectangula* and *Ilyocryptus brevidentatus* and two ostracods *Notiocypridopsis frigogena* and *Eucypris fontana* browse or scavenge on the benthic mats, which contain a microfauna of protozoa, turbellaria, rotifers, tardigrades, nematodes, gastrotrichs and enchytraeid annelids (Heywood and others 1979, Heywood, 1984).

Mesotrophic lakes. Heywood Lake typifies older, more mature lakes closer to the sea. It lies in two adjoining basins 4 m above sea level and 200 m from the shore, with a surface area of 45 000 m^2 and a mean depth of 2.0 m. An extensive catchment area with mineral soils, frozen in winter but waterlogged in summer, supports swards of moss and lichens which contribute minerals to the water (Table 4.2); the lake is also visited by seals in summer. Waters are clear under the ice in winter but turbid from organic and mineral particles when the ice has melted.

Salinity and nutrient values are very much higher than those of Moss Lake, and the waters are markedly more productive. Chlorophyll *a* values up to 170 mg m^{-2} were recorded in spring and up to 40 mg m^{-2} in summer. Over 3 g C m^{-2} were fixed daily during the summer peak, and the annual rate of production was estimated at 173.0 g C m^{-2} (Light *et al.*, 1981) or 139–270 g C m^2 (Priddle *et al.* 1986). Lakes that are rich in phytoplankton and other suspended matter tend to have correspondingly less benthic vegetation, presumably because of light limitation in summer. Bacteria were plentiful in the water column of Heywood Lake; Hawes (1985) provides evidence that they enhance algal photosynthesis in lakes where nutrients are low, by speeding the recycling of carbon, nitrogen and phosphates. Planktonic crustaceans include *Pseudoboeckella poppei, Parabroteas sarsi* and immature stages of the fairy-shrimp *Branchinecta gaini*; mature stages tend to browse deeper on the benthos. There are fewer invertebrate species in the benthic algal mats, which are less developed than in Moss Lake.

Eutrophic lakes. Amos Lake lies 10 m above sea level, adjacent to the shore on the windward side of the island; like many others in this situation, it is vastly enriched by marine influences. A small lake of surface area 6000 m and mean depth 1.7 m, it has a relatively small surrounding catchment basin, but the area is a popular summer resort for seals and nesting giant petrels *Macronectes giganteus.* There is little vegetation, but excreta and moulted hair contribute organic material to the soils, and the waters of Amos Lake are saline, turbid and nutrient-rich (Table 4.2).

Well stirred by seals, Amos Lake was too murky for direct productivity comparisons with the other lakes. Productivity was concentrated close to the surface; chlorophyll *a* values were high in summer and a daily peak production of over 300 mg C m^{-2} was recorded, decreasing rapidly below 1 m (Heywood, 1984). *Pseudoboekella poppei, Parabroteas sarsi* and *Branchinecta gaini* were present in the water column, but there was little or no benthic vegetation or associated microfauna. Bacterial counts were generally high. Both aerobic and anaerobic bacterial decomposition contributed to nutrient recycling in lakes at all nutrient levels (Ellis-Evans, 1982; Hawes, 1985); the role of anaerobic bacteria is enhanced in winter when the water is static under its ice cover. Where nutrients are plentiful, productivity is high, production far outstrips decomposition, and organic sediments accumulate on the lake beds.

Saline lakes. Lakes with densely saline waters are found in several of Antarctica's oasis areas. Their origins and histories are varied, but all occupy ground that was previously glaciated and many are ancient. Among the largest and most fully-studied lake systems, Lake Bonney in the Dry Valleys of South Victoria Land has probably existed for over 100 000 years (Hendy *et al.*, 1977), and a sediment core 2.8 m long from neighbouring Lake Fryxell is estimated from C-14 dating to represent over 50 000 years' deposition (Parker and Simmons, 1985). Similar lakes are known also in smaller coastal oases at Bunger Hills, Vestfold Hills, Schirmacher, Soya and Thala Hills. The ecology of saline lakes has been reviewed by Heywood (1984) and Vincent (1988).

Saline lakes of the Vestfold Hills oasis (Figure 4.5) string out along a valley, formerly a marine inlet, which rose and was cut off from the sea 5000–8000 years ago when the local land ice melted (Kerry *et al.*, 1977). Their salinities vary according to how long ago their waters were left to evaporate, and the amount of fresh water entering them annually; of the four main lakes, Deep Lake, the deepest (34 m) and most salty, has a mean salinity of approximately 280 ppt, roughly eight times that of sea water and similar in density to Dead Sea water. Deep Lake, and the slightly less salty Club Lake, seldom freeze over, though their surface temperatures drop almost to $-20°$ C in winter. Wind-stirred in all but the deepest layers throughout winter, in summer they become thermally stratified, with surface temperatures over 10° C. Algae and bacteria are present in all the lakes at all depths; productivity is very low, and few herbivores or scavengers have been recorded. Metazoans are virtually absent except in some of the lakes where fresh water seeps in at the edges.

In higher latitudes, the saline lakes and ponds of South Victoria Land present a wide spectrum. Many are well above sea level and appear to owe their high salinity entirely to long-term accumulation and evaporation (for a discussion of this point see Heywood, 1984). Don Juan Pond, 122 m above sea level in Wright Valley and now only a few centimetres deep, is assumed to be the remnant of a much larger freshwater lake of at least 10 m depth. Currently fed by a saline stream and limited snow-melt, its salinity, dominated by calcium chloride, is about 13 times that of sea water, and high enough to ensure that it seldom if ever freezes. The pond is virtually sterile except for algal mats which grow only at points about its edges where fresh water is available (Harris *et al.*, 1979; Wright and Burton, 1981).

Lake Vanda, also in Wright Valley, is a highly-stratified lake some 5 km^2 in area and up to 70 m deep. The remnant of a larger lake that once filled the surrounding valley, it has a superficial layer of fresh or slightly saline water down to 50 m deep, replenished seasonally by glacial streams, notably Onyx River, and stirred by their turbulence. Beneath lies a core of ancient salty water with a concentration of 140 ppt and a temperature of 25°C; the origin of this warmth is uncertain, but more likely to be solar than geothermal. The lake surface is permanently frozen to a depth of 2–4 m, with ice that thaws only around the edges in summer. Several other lakes in the area share some or all of these characteristics.

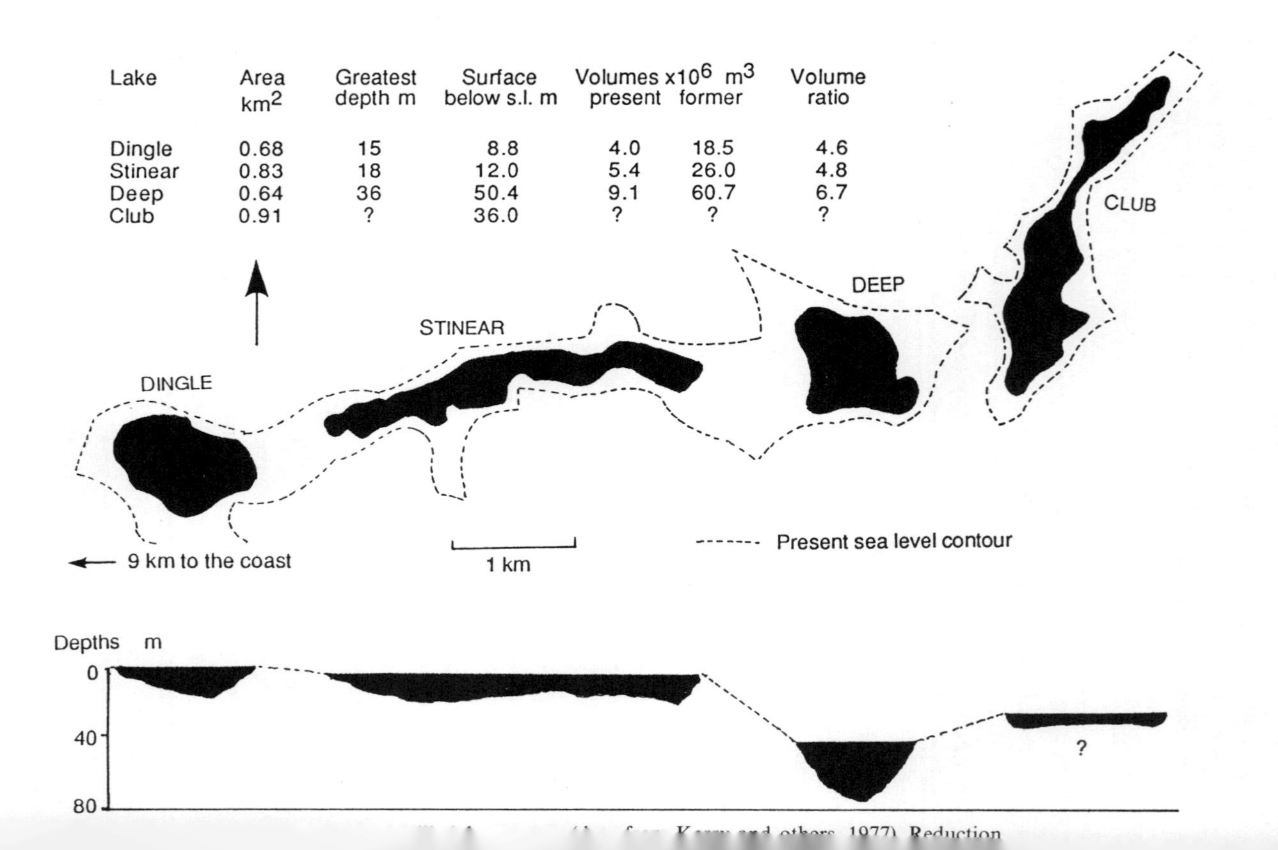

Lake	Area km^2	Greatest depth m	Surface below s.l. m	Volumes $\times 10^6$ m^3 present	Volumes $\times 10^6$ m^3 former	Volume ratio
Dingle	0.68	15	8.8	4.0	18.5	4.6
Stinear	0.83	18	12.0	5.4	26.0	4.8
Deep	0.64	36	50.4	9.1	60.7	6.7
Club	0.91	?	36.0	?	?	?

The ice cover inhibits wind-induced turbulence, minimizes penetration of sunlight and keeps the temperature of the surface waters almost constant. Algae are present in the upper layers; productivity is low even in summer, and probably restricted to a shallow near-surface layer beneath the ice. Oxygen produced by photosynthesis accumulates in surface waters, but the saline substratum is anoxic; nitrifying bacteria accumulate at the boundary between, forming a substrate where algae and heterotrophic bacteria also congregate, and photosynthesis is maximal. Sulphur bacteria may be present below the anoxic zone, reducing sulphate ions and liberating hydrogen sulphide. Benthos around the lake edge includes mats of blue-green algae and moss, often with a microfauna of rotifers, tardigrades and other scavengers.

In the productivity of Lake Vanda phosphorus is a limiting element (Vincent and Vincent, 1982), because of gravitational losses into the sediment and lack of turbulence to return it to general circulation. Phosphorus that diffuses upward from the bottom is absorbed by the accumulated bacteria and algal cells at the top of the anoxic layer; very little penetrates to the less saline layers above, which are thus limited to the very small amounts brought in by Onyx River. For a full discussion of nutrient cycling in this remarkable lake see Vincent (1988).

Running-water communities. Antarctica has many small seasonal streams, and a few large enough to be called rivers. Nearly all are coastal; those of the continent are fed mainly from melting glaciers, those of the maritime region mainly from melting snow. Usually dry throughout winter, they start with the onset of spring thaw. Unbuffered by variations in gradient, their flows often fluctuate widely from afternoon maxima to early morning minima. Their waters seldom have opportunities to spread and be warmed; temperatures are usually less than 5°C. Continental streams run for one to three months annually, those of the maritime Antarctic for four to six months or more. Many carry finely-divided glacial sediment.

Plankton in Antarctic streams is usually low, restricted to algae or other organisms washed from lakes or ponds along the route. Stream beds are sometimes more productive, those with clear water supporting films or mats of cyanobacteria (*Phormidium, Nostoc, Oscillatoria*) and epilithic algae (*Binuclearia, Prasiola*). Beds of continental streams dry out quickly after the brief summer flow; any ice present ablates, and the desiccated mat is subject to intense winter cold. Those of the maritime region remain moist for a longer season, and usually remain protected by ice or snow, if not by water, throughout winter. In consequence they support richer algal mats, often with well-developed filamentous species (for example *Klebsormidium, Mougeotia*) which are active for much of the year (Vincent, 1988). Though they appear green and contain abundant chlorophyll, productivity of these mats is generally low: much of their material is moribund, and persists only because of an absence of decomposers.

Though the periantarctic islands are well endowed with ponds, lakes and streams, comparatively little is known of their biology.

4.4 Summary and conclusions

Though water is a solid for much of the polar year, and generally poor in minerals even when in liquid form, both snow and aquatic polar habitats support life. Melting coastal snow supports an ephemeral algal flora that stains it red or green. Polar ponds and lakes are limited in their energy uptake by snow-covered ice, which also inhibits wind-stirring for much of the year. Permanently frozen catchment areas limit the amount of minerals available, though there is usually a small intake from soils during the summer thaw. Because food chains are simple, minerals absorbed into primary production settle in sediments where, in the absence of decomposers and detritus feeders, they may be lost to circulation. Some polar lakes, especially deep ones close to the coast and those in snow-free areas, may be saline with strongly-developed layering. Productivity is generally low in both still- and running-water habitats, limited mainly by lack of nutrients.

CHAPTER FIVE
MARINE ENVIRONMENTS

5.1 Introduction: polar seas

As mariners have long been aware, a voyage from tropical to cold oceans is a passage from dearth to plenty. Tropical oceanic waters, generally poor in nutrients and productivity, support only a fraction of the biomass found in the cold northern and southern oceans. Except inshore, and in areas of upwelling where local productivity is high, tropical seas have relatively few fish, seabirds, porpoises and whales. There are no congregations of tropical seabirds to match the huge concentrations in high latitudes; nineteenth- and early twentieth-century whaling and sealing were predominantly polar and subpolar activities, pursued despite the hazards of cold, stormy environments, and commercial deep-sea fisheries still thrive on the polar fringes.

That polar seas are richer than polar lands is immediately apparent at either end of the world. The arctic tundra supports both birds and mammals, at densities that are low in summer and even lower in winter; by far the densest concentrations of both are of marine species that live and breed ashore but feed entirely at sea. High arctic human populations are based on land, but usually live close to the sea and draw much of their sustenance from it. The antarctic continent has no indigenous land birds or mammals, and man cannot live off the land there; however, many millions of marine birds and mammals feed at sea and come ashore or onto the sea ice to breed. Subpolar seas of both hemispheres are richer than polar seas, with a longer but less intense season of high productivity.

From the concentrations of birds and mammals in high latitudes has grown the myth that all polar seas are immensely productive. On the whole they are not, but the seasonal productivity of a few areas is very high, and subpolar seas include some of the richest patches of ocean known anywhere in the world. The pack ice and fast ice that spread over polar seas form vast sheets many hundreds of square kilometres in area. They form a barrier that isolates the sea surface from the atmosphere and reflects away solar radiation, but they are also important biological habitats, supporting a

Table 5.1 Some similarities and differences between the Arctic and Southern oceans. Modified from Knox and Lowry (1977) and Hempel (1985).

Feature	Arctic Ocean	Southern Ocean
Area (10^6 km^2) of ocean	14.5[a]	37.5
Depth, continental shelf	100–500 m	400–600 m
Latitudinal range	70°–96°N	50°–70°S
Direction of currents	Transpolar	Circumpolar
Pack ice:		
seasonality	Low	High
area (10^6 km^2) and		
% mean cover, max	12.5: 85%	22: 60%
min	6.5: 45%	2.5: 7%
Icebergs and ice islands	Small, rare	Large, plentiful
River output	Massive	Negligible
Vertical mixing	Little	Much
Illumination	Strongly seasonal	Weakly seasonal

[a] Includes Barents Sea, Canadian Strait, Baffin Bay, Hudson Bay and Labrador Sea (Gierloff-Emden, 1982).

distinctive microbiota, providing shelter for planktonic animals, and forming a floating platform on which birds settle and seals breed.

The Arctic and Southern oceans differ in many respects; some key similarities and differences are summarized in Table 5.1.

5.2 Polar marine habitats

5.2.1 *Sea ice formation*

Sea ice starts to form when surface waters are at freezing point and turbulence is low. Ice crystallizes out and accumulates in a surface layer which thickens and consolidates; on calm nights with air temperatures well below freezing point a layer 20 cm deep may form overnight and be firm enough to take the weight of a man by morning. This amorphous frazil ice is easily broken by swell into small floes that rub against each other and continue to grow, developing characteristic upturned edges of pancake ice in a matrix of loose crystals (Eicken *et al.* 1988). In continuing calm and cold conditions the pancakes fuse together and the ice sheet thickens further, by growth of amorphous crystals and accretion of ice platelets—flat crystals—from below (Figure 5.1).

Once established, the pancakes and solid sheet ice are subject to lifting by swell, which may be generated many miles away in open water, and to lateral pressures from winds; sections split into floes which raft over each

Figure 5.1 Lead in young sea ice. This early-spring ice sheet, three to four weeks old, is formed of 20 cm of solid ice, overlain by 15 cm of blown snow. Wind-induced lateral pressure has caused ridging and splitting. The undersurface is a substrate for a yellow-brown layer of photosynthesizing diatoms, and for the zooplankton that feed on them.

other, or form pressure ridges along their edges. Rime and snow collect on the upper surface, depressing the floes and allowing the sea to flood over, freeze, and strengthen the ice further. During the growth period planktonic bacteria, diatoms and even minute animals are built into the ice, forming communities which develop and stain the floes, especially the undersurfaces, green or brown.

The ice crystals are pure water, and the concentrated brine that remains after their formation gathers in pockets and channels between the crystals, ultimately to drain back into the sea and sink away from the ice-sheet. On melting in spring the ice provides a thin layer of low salinity at the surface, and often an already-flourishing community of algae, which waves and wind-stirring soon disperse.

5.2.2 *Arctic seas*

The Arctic Ocean, including the Barents, Kara, Laptev, East Siberian, Chukchi and Beaufort seas, covers about 14 million km^2, almost two-thirds of the Arctic region as a whole (CIA, 1978). The deepest area is a central depression some 2500 km long and 1500 km wide, surrounding the North Pole and oriented at a right angle to Greenland (Figure 5.2). This is crossed by a submerged mountain range, the Lomonosov Ridge, which runs between northern Greenland and the New Siberia Islands (Ostrova Novosibirskiye), flanked by a lesser ridge on either side. The Lomonosov Ridge rises from a sea-bed some 4000 m deep to within about 1000 m of the surface, splitting the depression into Eurasian and Amerasian basins; the

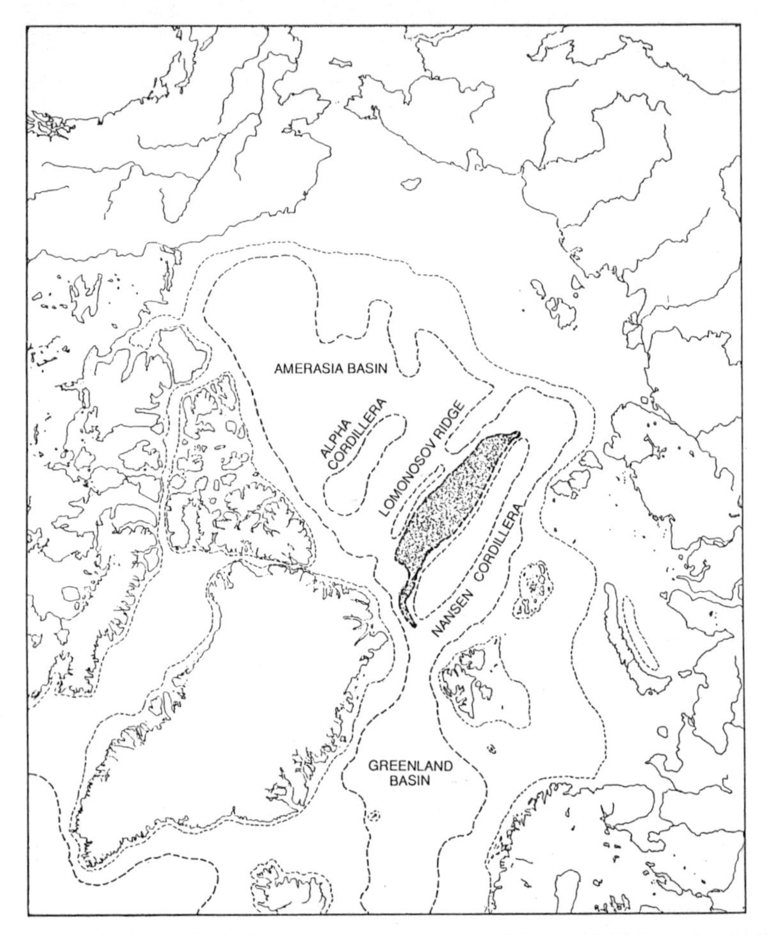

Figure 5.2 The Arctic Basin. Submarine contours are at 200 m and 2000 m; the shaded areas are below 4000 m.

Eurasian basin is the deeper, showing soundings of 4500 m between northern Greenland and the Pole. A single deep channel links the Eurasian basin with the submarine Greenland Basin east of Greenland; a shallower channel links it with the Baffin Basin west of Greenland.

The continental shelf surrounding the depression is less than 200 km wide off Greenland and North America but 500–1000 m wide off Eurasia; altogether it totals about half the area of the ocean. North of Greenland and central Canada the shelf is down to 450 m deep. Under the Barents Sea between Scandinavia, Svalbard, Severnaya Zemlya and Novaya Zemlya it is 200–350 m deep, and off most of Siberia, extending eastward to Alaska, it forms a broad submarine plain only 60–100 m deep.

Arctic Ocean waters are continuous with the Pacific Ocean via the 80-km wide Bering Strait, and with the north Atlantic Ocean on either side of Greenland, via the 1400-km wide Norwegian Sea in the east and the much narrower channels of the Canadian archipelago in the west. A net inflow of about 8000 km^3 of warm salt water enters the basin annually through the Bering Strait. A very much larger flow, possibly 400 000 km^3, enters annually from the Norwegian Sea (Kort, quoted in Zenkevitch, 1963). Compared with these flows, inputs of fresh water from the rivers of North America and Asia are negligible, though their local effects are marked especially in summer (Chapter 4). Balancing outflows of about 430 000 km^3 of water and 6000 km^3 of ice pass southward, mostly along the eastern flank of Greenland but also through the Canadian islands.

Arctic sea ice. Winter sea ice covers some $12–13 \times 10^6$ km^2 of the Arctic basin, extending over most of the ocean surface; the amount of sea ice varies from season to season, but usually only the southwestern Barents Sea north to Murmansk remains open to shipping. Sea ice also covers Baffin Bay. Hudson Bay, much of the Labrador Sea and the St Lawrence estuary, invests the coasts of Labrador, Greenland, Svalbard, western Alaska and eastern Siberia, closing the northern Bering Sea and filling the Sea of Okhotsk as far south as the Kuril Islands and northern Japan.

Summer cover in the Arctic basin is reduced to a central asymmetrical core of 5–8 million km^2, about half the winter area (Figure 5.3). The summer ice presses against Greenland and the eastern North American coast, which are thus effectively closed to shipping all the year round. From about June onward there are navigable waterways through the Barents, Kara and Bering seas and along much of the Siberian coast (the 'northeast passage'), though heavy pack ice usually persists between the mainland and Severnaya Zemlya and off Wrangel Island. There are also wide coastal

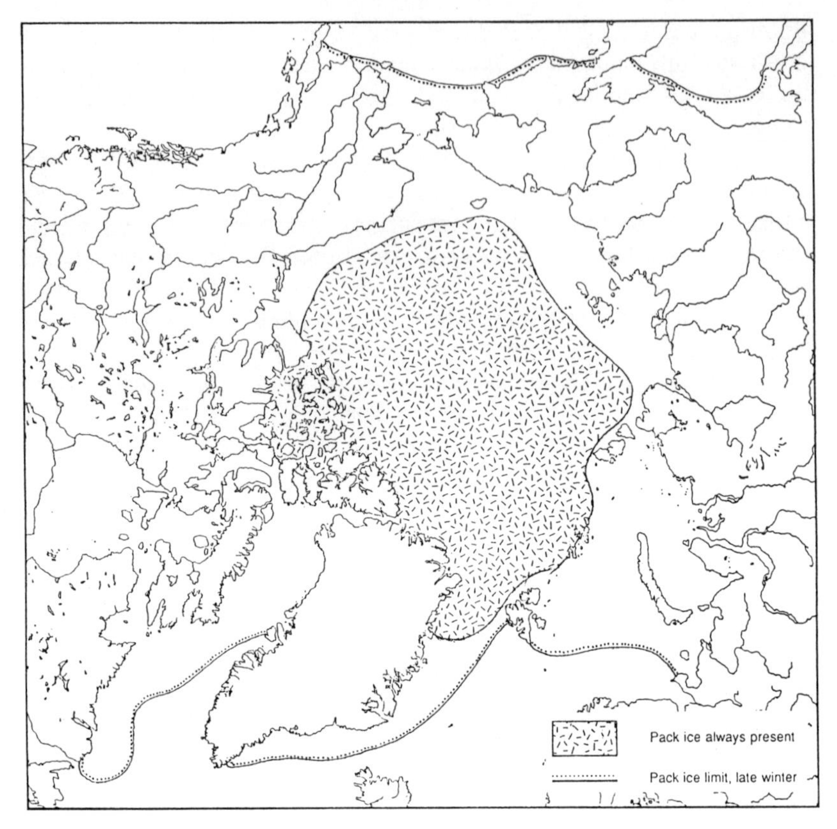

Figure 5.3 Year-round and late winter extent of Arctic sea ice.

lanes along the northern shore of central and western North America. Outside the basin only northeast Greenland and the channels of the Canadian archipelago remain heavily invested in ice, though a 'northwest passage' is usually navigable between the mainland and Banks and Victoria islands.

Even in the depths of winter the sea ice is never a complete cover. Slow clockwise circulation of the ice mass in the Arctic basin, coupled with local winds and both horizontal and vertical (upwelling) currents, cause the formation of wide cracks and polynyas (anomalous patches of open water). These openings, some of which occur in the same place year after year, are often of great biological importance, allowing light to penetrate, phytoplankton to bloom, and whales, seals and seabirds to survive when the rest of the sea remains covered. During summer loose pack ice from the Arctic Ocean streams southward through the Greenland Sea and down the east Greenland coast; lesser streams pass from the Barents Sea to the Norwegian Sea and from Baffin Bay down the Labrador coast, ultimately to melt in the warmer waters of the north Atlantic Ocean.

Icebergs and ice islands. Interspersed with ice floes are larger masses of ice that have broken from glaciers and ice shelves on land; icebergs are the fragmented products of glaciers, ice islands the larger masses, up to 60 m thick and 30–40 km long, broken from ice shelves. Both are produced from the ice sheets of Ellesmere, Devon and Baffin Islands and Greenland (Jeffries, 1987). Some circulate in the Arctic Ocean, others stream down the east Greenland or Labrador coasts and disperse widely in the north Atlantic Ocean, becoming a hazard to shipping. A few reach 40°N before melting away; a very few are large enough to continue southward, occasionally as far as New York and Bermuda. Like ice floes, small bergs and ice islands form roosts for seals and seabirds, and often carry an interesting microbial flora and fauna.

5.2.3 *Antarctic seas*

The Southern Ocean is a broad, deep ring of water, contiguous with the southern Atlantic, Indian and Pacific oceans but bounded by the Antarctic Convergence (p. 11). Its total area of about 28 million km^2 (roughly twice that of continental Antarctica) includes many small marginal seas and two much larger ones, the Weddell and Ross Seas, occupying shallow bights on either side of the continent. The continental shelf lies unusually deep because of the vast burden of ice on the land, mostly between 400 and 800 m below sea level. Except in the Ross and Weddell bights and along the coast of West Antarctica the shelf is generally narrow, rarely more than 100 km wide.

From the shelf edge the continental slope descends steeply to an abyssal plain, partitioned into deep Atlantic, Indian and Pacific ocean basins by submarine ridges and platforms (Figure 5.4). Mostly between 3000 and 5000 m deep, the sea-bed falls to over 8000 m in an abyssal trench east of the South Sandwich Islands. A submarine mountain arc links South Georgia, the South Sandwich, South Orkney and South Shetland Islands and Antarctic Peninsula. Iles Crozet and the Prince Edward Islands share a submarine platform, and a much shallower platform links Iles Kerguelen and Heard Island.

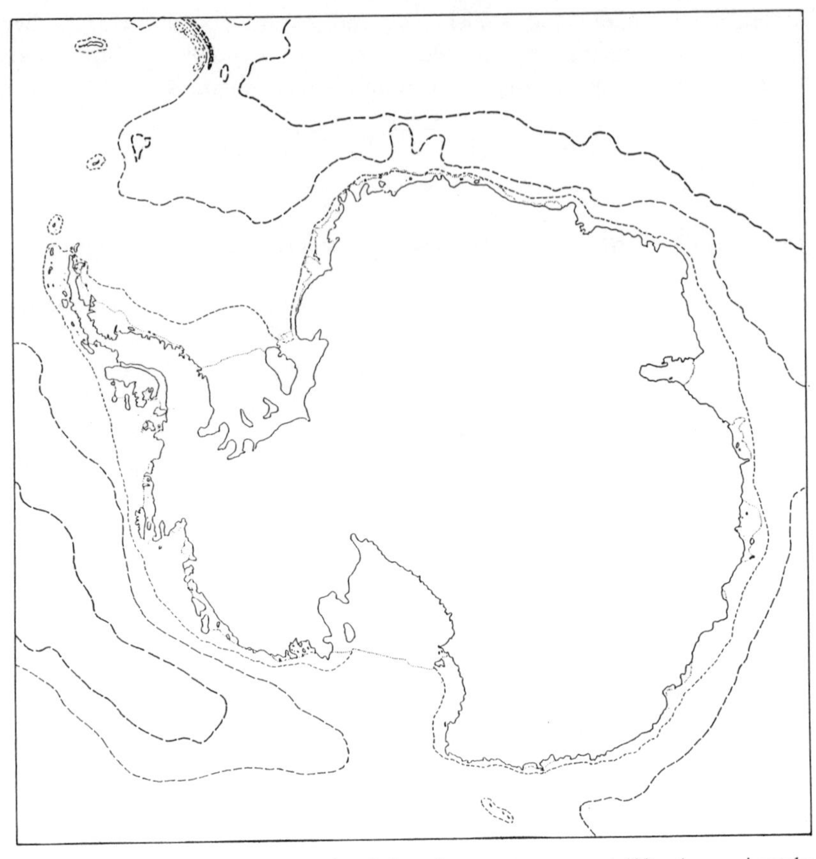

Figure 5.4 The seas around Antarctica. Submarine countours are at 400 m (approximately the edge of the depressed continental shelf), 4000 m and 5000 m. The shaded area (South Sandwich Deep) descends to over 8000 m.

Antarctic sea ice and icebergs. A major feature of the ocean is the ring of shifting pack ice that occupies the southern half, permanently surrounding Antarctica and increasing four- to fivefold from summer to winter. In autumn fast ice grows out from the land to meet new ice that forms at sea; from a March minimum of 3–5 million km^2 the pack ice spreads to 17–20 million km^2 by September (Foster, 1984), an annual variation far greater than that of the Arctic Ocean (Figure 5.5). Despite its persistence, much of the pack at any time is less than one year old, averaging 1–2 m thick. Only in corners where fast ice fails to break out each summer, or where pack

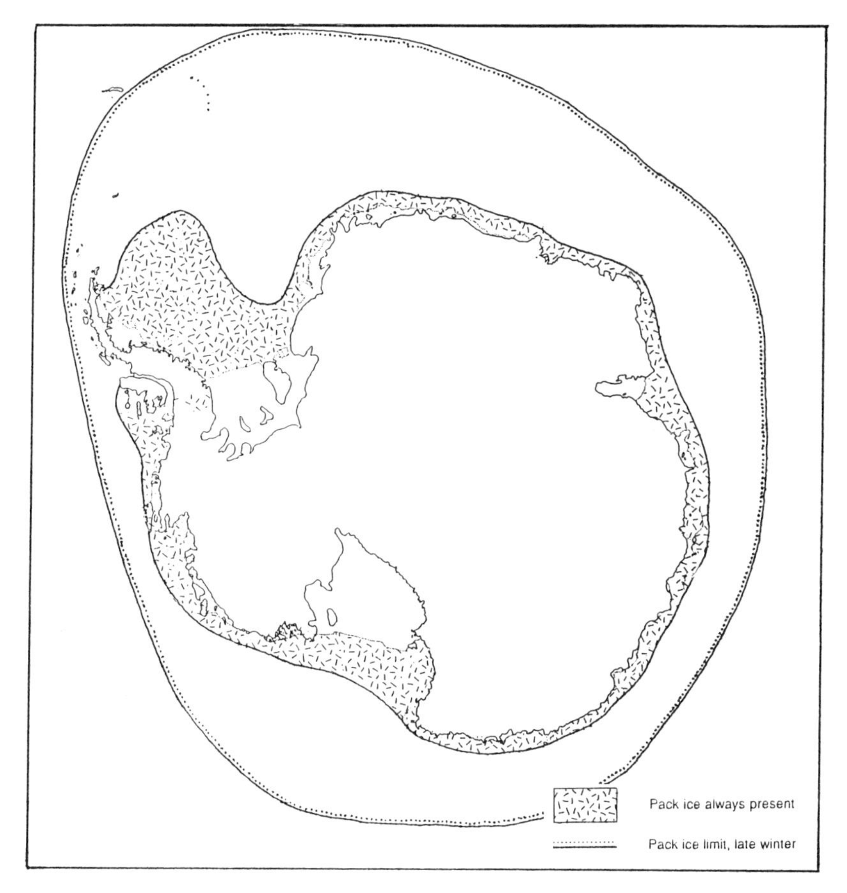

Figure 5.5 Year-round and late-winter extent of Antarctic sea ice.

accumulates and circulates in local gyres, do the floes last several years, and thicken both by pressure-rafting and by accretion. The western Weddell Sea, the eastern Ross Sea and much of the Pacific coast of West Antarctica are notorious for their heavy, multi-year pack ice.

The September–October northern limit of pack ice varies considerably from year to year. Though often hazy and irregular due to wind scattering, the edge is visible from satellites and can be monitored month by month (Foster, 1984). In some winters it extends as far north as the limit of cold Antarctic surface waters, i.e. to the Antarctic Convergence, particularly in the Pacific and Atlantic sectors of the ocean. Mean monthly positions of the

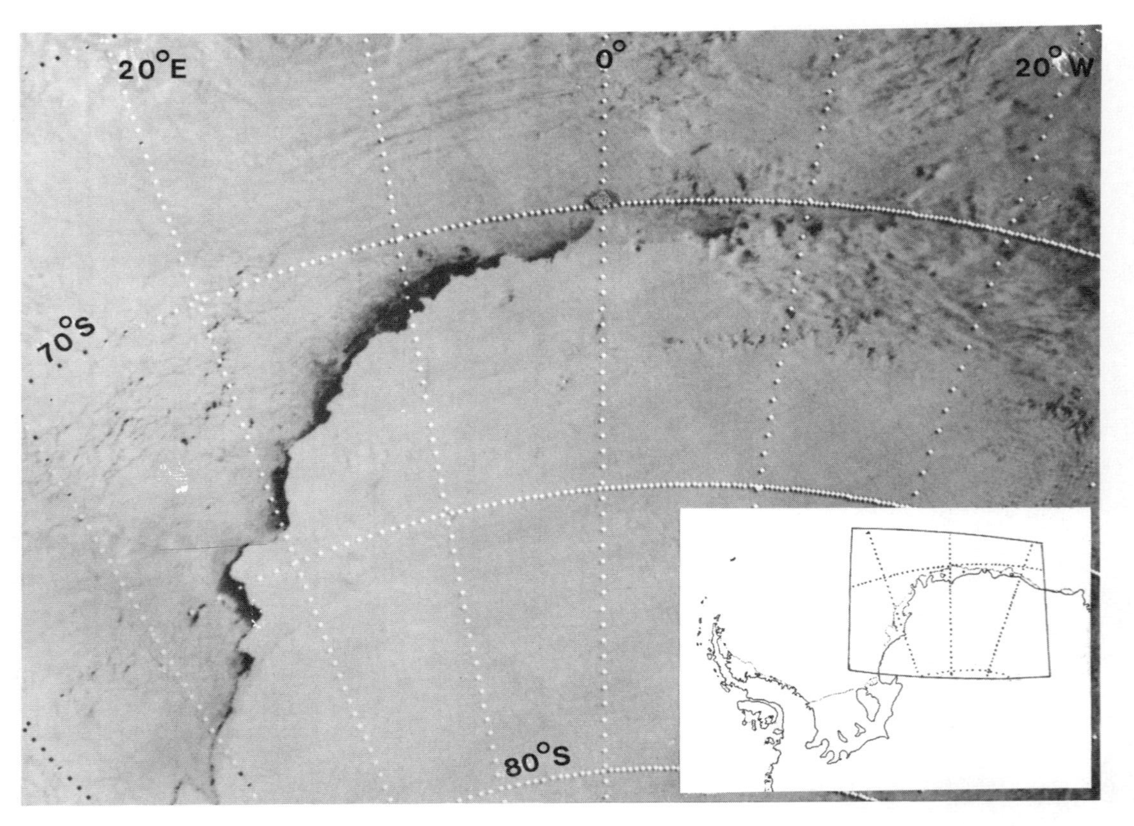

Figure 5.6 Satellite image of part of the Weddell Sea coast of Dronning Maud Land and Coats Land, Antarctica, in early spring, showing a coastal polynya (dark) extending from 71°S to 77°S. Patches of open water can be seen among the pack of ice of the Weddell Sea; cloud obscures the northeast corner. Inset: the area covered.

ice edge are now well established for all sectors of the Southern Ocean (Bakayev and others, 1966).

Within the pack ice polynyas and leads form and re-form constantly (Eicken *et al.*, 1988), creating openings that are biologically no less important than those of the Arctic (Figures 5.6, 5.7). The presence of coastal polynyas in high latitudes in early spring, for example, allows seals and seabirds to start breeding before the main ice-sheet has dispersed; indeed the presence of a breeding colony on the Antarctic coast is a sure indicator of the presence of open water in early spring, before the main ice-sheet has dispersed (Stonehouse, 1967). Polynyas are important also in giving man access to high latitudes (Figure 5.8).

Flotillas of tabular icebergs and ice islands sail among the pack ice. Generated from the antarctic ice-sheet, they are far more numerous and often much larger than their arctic counterparts, the largest exceeding 200 km long when newly formed. Three ice islands recently broken out from the Ronne-Filchner ice shelf, in the southeast corner of the Weddell Sea, in area totalled 11.5 thousand km^2 and in thickness some 500 m (Ferrigno

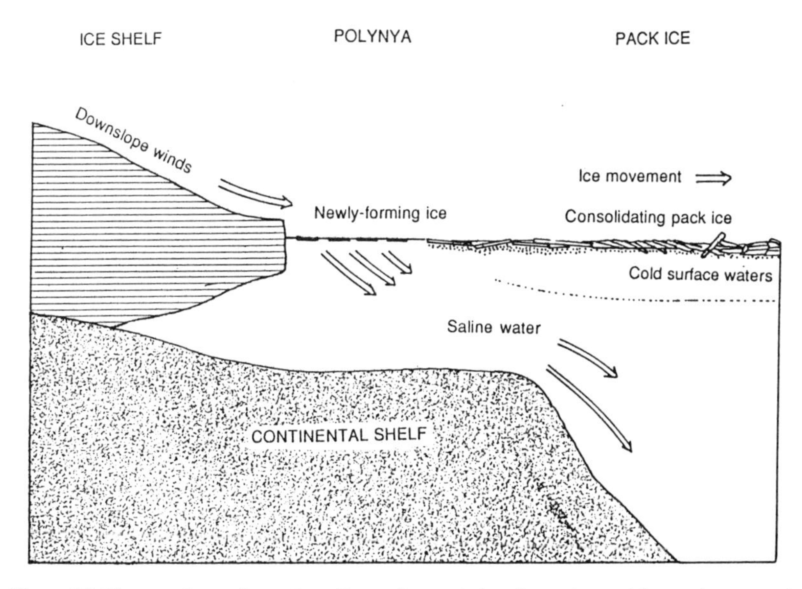

Figure 5.7 Shore polynya formation. Downslope winds, often triggered by cyclones, push ice away from the coast, consolidating it by ridging and rafting. In winter and early spring new ice is constantly being formed, a residue of highly saline water flowing down the continental slope to form Antarctic Bottom Water (see Figure 5.11). Thin, translucent ice early in spring is an important substrate for diatom growth.

Figure 5.8 West German research icebreaker *Polarstern* deploying scientists on newly-formed ice close to the Weddell Sea polynya. Photo: B. Stonehouse.

and Gould, 1987); a single one almost 160 km long with an area of over 6200 km^2, released from the Ross Ice Shelf in 1987, was estimated to contain enough water to last the State of California over 600 years (*Polar Record*, 1988). Movements of large tabular bergs have been monitored from satellite photographs (Swithinbank and others, 1977) or by signals from radio transmitters tracked via satellites (Tchernia and Jeannin, 1984). They tend to follow recognized paths westward through the zone of easterly winds immediately surrounding the continent, then eastward and north through the westerlies. The largest pass beyond the pack ice limit and out of the Southern Ocean altogether, melting, fragmenting, and disappearing in the warm subantarctic waters beyond.

5.2.4 *Food webs*

Food webs are basically similar in northern and southern polar seas (Figure 5.9). Primary producers, as in all other oceans, are the phyto-plankton—single-called plants of surface waters. Almost moribund in winter, the cells proliferate from early spring onward, starting even before the dispersal of the sea ice and continuing through late spring and summer for as long as supplies of nutrients last. Diatoms of diameter greater than

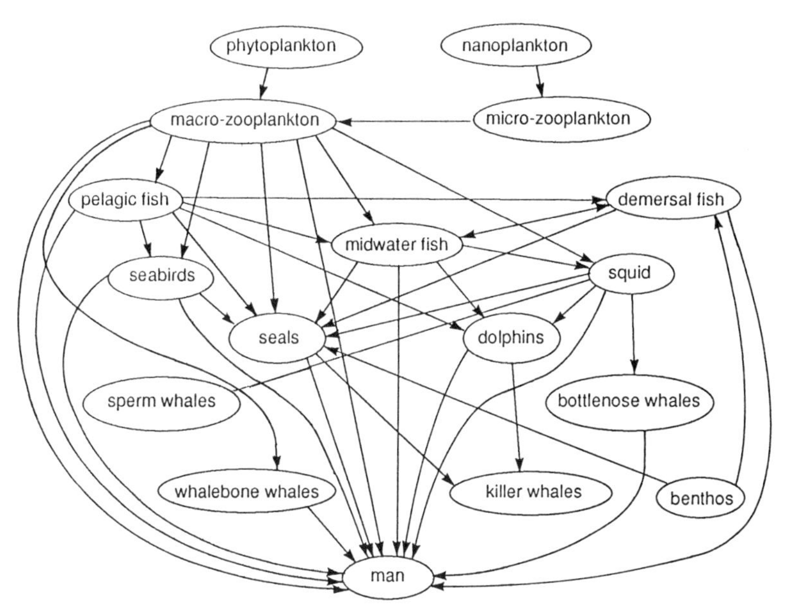

Figure 5.9 Food web for polar seas. Note: Most seals show degrees of specialization, hunting at different levels; for example walruses and elephant seals feed deep, respectively on molluscs and squid, while crabeaters and fur seals feed pelagically on krill.

35 μm are a prominent component of phytoplankton at either end of the earth, and until recently were thought to provide most of the primary production. There is now increasing evidence that smaller cells of diameter 2–20 μm have been seriously underestimated in energy budgets; they may be responsible for half the total primary production in southern waters and no less in the north.

Zooplankton includes herbivores that browse exclusively on phytoplankton, carvnivores that feed on neighbours, and omnivores that take impartially plant, animal or detrital material. Crustaceans, notably copepods, form the bulk of the browsers: euphausiids (including 'krill') are especially prominent in antarctic waters, where they form the main food of many birds and mammals. Larvae of fish and bottom-living invertebrates, which are often prominent in zooplankton of temperate seas, are notably absent from polar surface waters. Bacteria, protozoa and other micro-scavengers are likely to play important but as yet uninvestigated roles in the zooplankton.

A few species of pelagic fish feed on zooplankton, but the major predators in surface waters of either polar region are sea birds, mysticete whales and pelagic seals. Other species of fish, whales and seals feed in mid-water or on the sea-bed, where their prey are mostly fish and squid. Both polar regions have patchy and rather limited stocks of benthic animals; of the two, the Antarctic is far richer in both variety of species and numbers of individuals.

5.3 Arctic Ocean ecology

5.3.1 *Northern water masses*

Coachman and Aagaard (1974) distinguish four major arctic water masses. The top 50–100 m of ocean forms the polar mixed layer, a layer diluted by fresh water (mostly from rivers; see below) to salinity of about 32 parts per thousand (ppt). In constant contact with ice, its temperature remains close to freezing point. Beneath lies a mass of slightly warmer, more saline water some 600 m deep. This is Atlantic water, which originates in the north Atlantic Ocean at a temperature of 2°C and salinity of 35 ppt, and cools almost to 0°C as it circulates slowly about the ocean basin; its salinity remains high at 34.9 ppt. Between these two lies a pycnocline layer, 100–150 m thick, of intermediate temperature, salinity and density, spreading from the continental shelf where it is formed during the freezing of surface ice. Filling the central depression is a layer of arctic deep water of temperature −0.9° to −0.5°C and salinity 34.94°C; it is slightly warmer in the European basin than in the Canadian, and probably of Atlantic origin. The water masses are illustrated in Figure 5.10.

The surface polar mixed layer circulates clockwise, driven by anticyclonic winds. Ice floes circulate with it across the Pole from eastern Siberia toward Ellesmere Island, Greenland, and out into the north Atlantic Ocean, mostly along the east Greenland coast. Deeper water layers circulate counter-clockwise, driven by the constant inflow from the Norwegian Sea along the Siberian shelf. Near Bering Strait the small inflow of Pacific water circulates locally in the pycnocline layer. Precipitation over the ocean is low, in volume probably exceeded five or six times by the fresh water that pours in from the rivers (Rudels, 1987), locally diluting the surface layer. Evaporation is low, but much fresh water is exported annually as ice floes 2–3 m thick.

5.3.2 *The plankton*

Phytoplankton. Because of its inaccessibility, the marine ecology of the Arctic basin is poorly known; most studies have been made from isolated

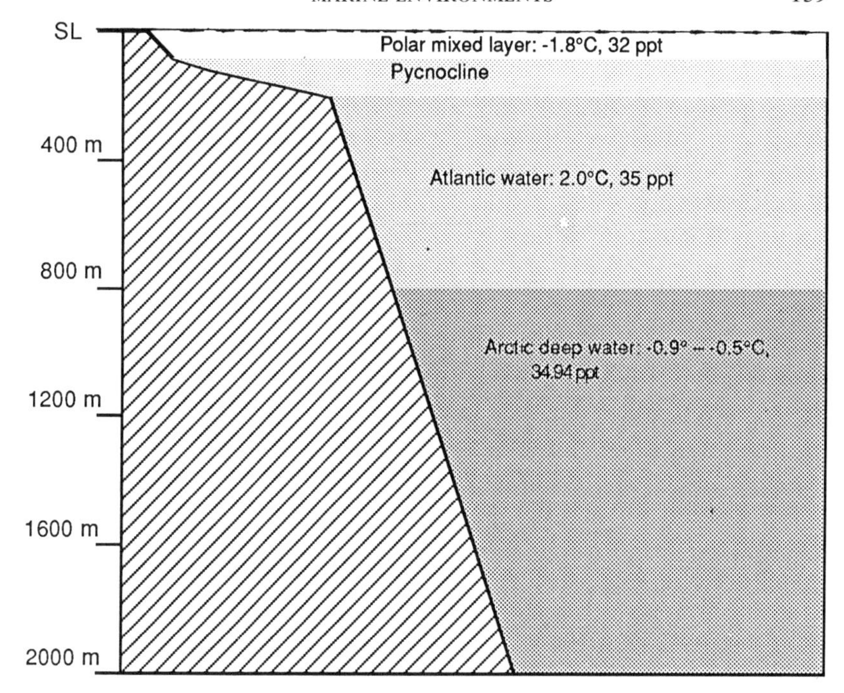

Figure 5.10 Arctic water masses.

shore stations off Alaska, Canada or Greenland (see for example Alexander, 1974; Bursa, 1961; Cairns, 1967, Clasby *et al*., 1976; Grainger, 1979), or from temporary drifting stations on ice floes manned by Canadian, U.S. or Soviet biologists (Brodsky and Pavshtiks, 1976; Dunbar and Harding, 1968, Zenkevitch, 1963).

Primary production is based mainly on diatoms, living in the ice and in open water. The season of production lasts probably less than two months in the Arctic basin, and up to four months in the peripheral region (read Bursa, 1961). Published estimates of primary production are generally very low; values of annual production of carbon per square metre, tabled by Grainger (1979), range from less than 1 g in the Arctic Ocean to 10–100 g off West Greenland and in channels and fjords among the Canadian islands.

Zooplankton. Because of the steady flow of Atlantic Ocean surface waters into the Arctic basin, most zooplankton species found there are of Atlantic origin. For example, of the 60 or more species of copepods found in arctic plankton, some 50 are of Atlantic origin; only a very few species penetrate from the Pacific, though some of those have been reported close to the Pole (Brodsky, 1956). Surface waters in contact with melting ice tend to be very thinly populated with zooplankton. There are consequently very few pelagic fish or diving birds to be seen, and the zone of permanent pack ice gives an impression of lifelessness. However, zooplankton is sparsely distributed at all levels.

From 30 m to 50 m down, the small copepod *Oithona similis* becomes plentiful; below this *Oithonia* and another small species, *Microcalanus pygmaeus*, are joined by the larger calanoid copepods *Calanus finmarchicus*, *C. hyperboreus* and *Metridia longa*. Other forms reported include rhizopod and radiolarian protozoans, hydromedusae, ctenophores, ostracods, amphipods, pteropods, chaetognaths and appendicularians (Brodsky and Pavshtiks, 1976). The richest and most varied zooplankton occurs below about 800 m in warmer Atlantic water.

Peripheral waters among the seasonal pack ice are richer in both species and individuals; the Barents Sea (Norderhaug *et al.*, 1977) and polynya areas among the channels of the Canadian Arctic archipelago (Stirling and Cleator, 1981) have specially favorable conditions of mixing and upwelling that promote relatively high productivity in summer. Arctic char *Salvelinus alpinus*, which spawn in rivers and fatten in the sea during their fourth and fifth years, capelin *Mallotus villosus*, which winter in deep water but rise to form huge surface shoals in summer, and arctic cod *Boreogadus saida* are among the few pelagic, plankton-feeding fish in both polar and subpolar waters. In deeper waters are sculpins (Family Cottidae), eel-pouts (Zoarcidae), polar cod *Arctogadus glacialus*, polar halibut *Reinhardtius hippoglossoides*; further south these give way to the more familiar cold-water species—northern cod, haddock, coalfish, skate, halibut, herrings, etc., that are hunted commercially for European and North American markets.

5.3.3 *Mid-water, sea-bed and shore*

Mid-water regions of oceans are always the hardest to explore and model; very little is known of this region of arctic waters.

Practically all arctic and subarctic shores are ice-bound from late autumn onward. Once air temperatures have fallen well below freezing point, frozen spray accumulates along the shore. When the sea freezes, a series of working tide-cracks form at the junction of land ice and sea ice, renewed twice daily during each tide cycle. The shore remains frozen until the summer thaw; during summer the beaches and rocky shores alike are scoured by floating ice; in consequence few plants or animals inhabit the intertidal or upper littoral zones.

The sublittoral is similarly scoured by shifting ice, but mats of short-stemmed brown and red algae flourish from about 2 m downward, below the level of ice floe movement. These grow actively from spring onward despite the persistent sea ice above; though little light penetrates the snow-covered ice, lack of wind-induced turbulence keeps the water clear. Close inshore the sublittoral flora and fauna is often disrupted by the formation of anchor ice, which grows on the sea-bed as heat is withdrawn from the land during the coldest months. Anchor ice accumulations, being less dense than sea water, break away from time to time and rise to the surface, carrying with them entrapped and frozen plants and animals, which gather in layers under the inshore floes.

The continental shelf has a covering of fine muds, derived mainly from the great continental rivers, and siliceous diatom shells from the plankton. Seaweeds grow to depths of about 100 m; below that level plants are rare, but the sea-bed is thinly carpeted with animals that live in or on the mud and feed on the rain of debris from above. Some feed by direct ingestion, for example nemertine worms, crabs, shrimps, starfish and brittlestars; others are filter-feeders that spread nets or pump water to filter minute particles of food, for example sponges, polychaete worms, euphausiid shrimps and clams. Demersal fish browse this carpet; the sea-bed in high latitudes supports many more species of fish than surface waters.

5.3.4 *Birds and mammals*

Seabirds. About 50 species of seabirds feed in arctic and subarctic waters, some in enormous flocks. Many, for example the auks and petrels, feed at sea all the year round, often at great distances from land; these are the most numerous, breeding in cliff colonies that include thousands of nesting pairs. Others, for example the cormorants, gulls and terns, are more coast-bound, feeding offshore in winter but mainly in shallow coastal waters in summer; they breed in smaller colonies, usually of a few dozen pairs close to good feeding grounds. Skuas and jaegers—predatory gull-like birds—winter at sea but feed almost entirely ashore on insects and small mammals during the breeding season; phalaropes, divers and several species of ducks similarly switch from marine foraging in winter to feeding and breeding on the tundra wetlands in summer. For notes on the size, distribution and feeding of arctic seabird communities see Croxall *et al.*, (1984) and Croxall

(1987); for reviews of their general biology see Stonehouse (1971) and Sage (1986).

Auks, guillemots and razorbills of the family Alcidae are the characteristic seabirds of the Arctic. Narrow-winged, small and compact, they fly powerfully with rapid wing-beats, sometimes planing over the water like hydrofoils; in hunting they dive and use their wings for propulsion under water, feeding mostly on plankton and small fish (Nettleship and Birkhead, 1985). Some 14 species breed in northern waters; those that range farthest north include Brunnich's guillemot *Uria lomvia*, smaller black and pigeon guillemots *Cepphus grylle*, which occur in the polar basin, and *C. columba*, common guillemots *U. aalge*, razorbills *Alca torda* and puffins *Fratercula arctica*, which are cliff-breeders, breed on many ice-bound islands. The garefowl or great auk, largest of the family and a non-flyer, bred in southern Labrador, Greenland, Iceland and north Britain. Constant harassment and predation on their breeding grounds by fishermen reduced their stocks during the eighteenth and nineteenth centuries; the last individuals were killed by man in 1844.

Only one species of petrel, the northern fulmar *Fulmarus glacialis*, breeds throughout the Arctic, on cliffs in Iceland, Svalbard, Greenland, the Canadian archipelago and Siberian high arctic islands. Gull-like in appearance, with gliding and soaring flight, they feed by settling on the water and bobbing for plankton at the surface. Atlantic storm petrels *Hydrobates pelagicus* and Leach's storm petrel *Oceanodroma leucorhoa* are temperate species that breed within the arctic fringe only in Iceland and Labrador. They nest mainly on rocky islets and feed by hovering and dancing over the water. Shearwaters visit subarctic feeding grounds in summer, Manx shearwaters *Puffinus puffinus* from nesting in Europe, sooty and slender-billed shearwaters *P. griseus* and *P. tenuirostris* from islands in the southern oceans. These dive and swim underwater for planktonic food, the partly-folded wings acting as hydroplanes that help the shearwaters to hunt below the level of superficial feeders.

Seals and whales. Of the world's 32 living species of seals, ten occur in or close to arctic waters (Stonehouse, 1985a; see data in Table 5.2.) Seals are aquatic mammals that feed at sea and emerge from it to rest, give birth and tend their pups. For these purposes sea ice is a useful extension of land. Several species live where sea ice is present; a few live almost entirely among pack or fast ice.

Life cycles and general biology of arctic stocks are summarized in King (1983). Of the three families of the order Pinnipedia, walruses and true or

Table 5.2 Polar seals.

	Weight (kg) and length (m) of adult male		Geographical range
Arctic species			
Walrus (*Odobenus rosmarus*)	1500	3.5	Inshore sea ice and islands, N. Canada, Greenland, Iceland, Kara and Barents seas
Ribbon seal (*Histriophoca fasciata*)	90	1.9	Inshore ice, N. Pacific, Sea of Okhotsk, Bering Seal, N. USSR
Harp seal (*Pagophilus groenlandicus*)	180	1.8	Offshore ice, Greenland and Barents seas, Labrador
Grey seal (*Halichoerus grypus*)	270	2.5	Iceland, Baltic, St Lawrence and temperate areas south
Bearded seal (*Erignathus barbatus*)	400	2.8	Circumpolar Arctic, inshore ice, N. Atlantic and Pacific
Ringed seal (*Pusa hispida*)	130	1.5	Circumarctic pack ice, Bering Sea, N. Pacific and Atlantic
Hooded seal (*Cystophora crystata*)	400	3.2	N. Atlantic and Arctic Ocean pack ice
Northern fur seal (*Callorhinus ursinus*)	270	2.2	Pribilov, Commander, Robben Islands
Antarctic species			
Weddell seal (*Leptonychotes weddelli*)	400	2.8	Circumpolar, continental inshore ice, fringe islands, S. Georgia
Ross seal (*Ommatophoca rossi*)	200	2.1	Circumpolar pack ice
Crabeater seal (*Lobodon carcinophagus*)	300	2.9	Circumpolar pack ice south to continent
Leopard seal (*Hydruga leptonyx*)	270	3.0	Circumpolar, N. edge Antarctic pack ice, coast of Antarctica, peripheral islands
Southern elephant seal (*Mirounga leonina*)	4000	6.0	Peripheral and cold temperate islands
Kerguelen fur seal (*Arctocephalus gazella*)	120	1.8	Antarctic fringe and peripheral islands

phocid seals are best represented in cold waters. Geographically separate stocks of some species are recognized, for example in walruses there is an Atlantic stock spread between Hudson Bay and the Barents Sea, and a Pacific stock found in the Bering, Chukchi, East Siberian and Laptev seas. Of the phocids, ringed, bearded, harp and hooded seals have the northernmost distributions, while ribbon and grey seals penetrate the arctic fringe, respectively in the Pacific and Atlantic oceans. Fur seals and sea lions are restricted to the Bering Strait area and seldom enter icy waters.

Seals take much of their food close to the surface, plankton and pelagic fish forming their main prey. Walruses and bearded seals, the two largest species, feed mainly on the sea bed, diving in shallow waters to depths of a few hundred metres, using their vibrissae to hunt in the mud for molluscs and crustaceans.

Killer whales *Orcinus orca* are the main predators of most species. Polar bears hunt several species, notably ringed seals, on the sea-ice breeding grounds, and man is a major predator of harp and hooded seals, valuing especially the pelts of pups and immature animals for commercial sales. Northern fur seals have for centuries been exploited commercially on the Pribilov and other islands in the Bering Sea; since 1911 stocks have been managed and exploition limited under international agreement, providing a rare example of successful management of an international resource (Young, 1981; Gulland, 1974).

Arctic fringe seas are well stocked with whales and dolphins—both toothed whales feeding on fish and squid, and baleen whales feeding mainly on small fish and krill. In far northern waters narwhal *Monodon monoceros*, beluga *Delphinapterus leucas* and Greenland right (bowhead) whale *Balaena mysticetus* have a circumpolar distribution along the ice edge and in loose pack. Krill-feeding right whales were heavily hunted by British, Dutch and other commercial whalers for baleen and oil during the nineteenth century (Brownell *et al.*, 1986); like narwhals and belugas, their stocks are now protected, and only limited non-commercial hunting by northern natives is allowed. Predatory killer whales and several other species of dolphins, northern beaked whales *Berardius bairdii* and northern bottlenose whales *Hyperoodon ampullatus* hunt in cold waters to the ice edge and beyond; summer visitors to the Arctic include grey whales *Eschrichtius gibbosus* and northern stocks of humpback whales *Megaptera novaeangliae*, male sperm whales *Physeter catodon*, blue whales *Balaenoptera musculus*, fin whales *Balaenoptera physalus*, sei whales *Balaenoptera borealis* and minke whales *Balaenoptera acutorostrata*.

5.4 Southern Ocean ecology

5.4.1 *Southern Ocean water masses*

Three main water masses can be distinguished within the Antarctic Convergence (Knox, 1970). Antarctic surface water forms a layer 70–200 m thick, ranging in temperature from freezing point (−1° to −1.8°C) to 3–4°C and in salinity from almost fresh to 34.5 ppt, or slightly higher over the continental shelf. This water originates at the Antarctic Divergence, a

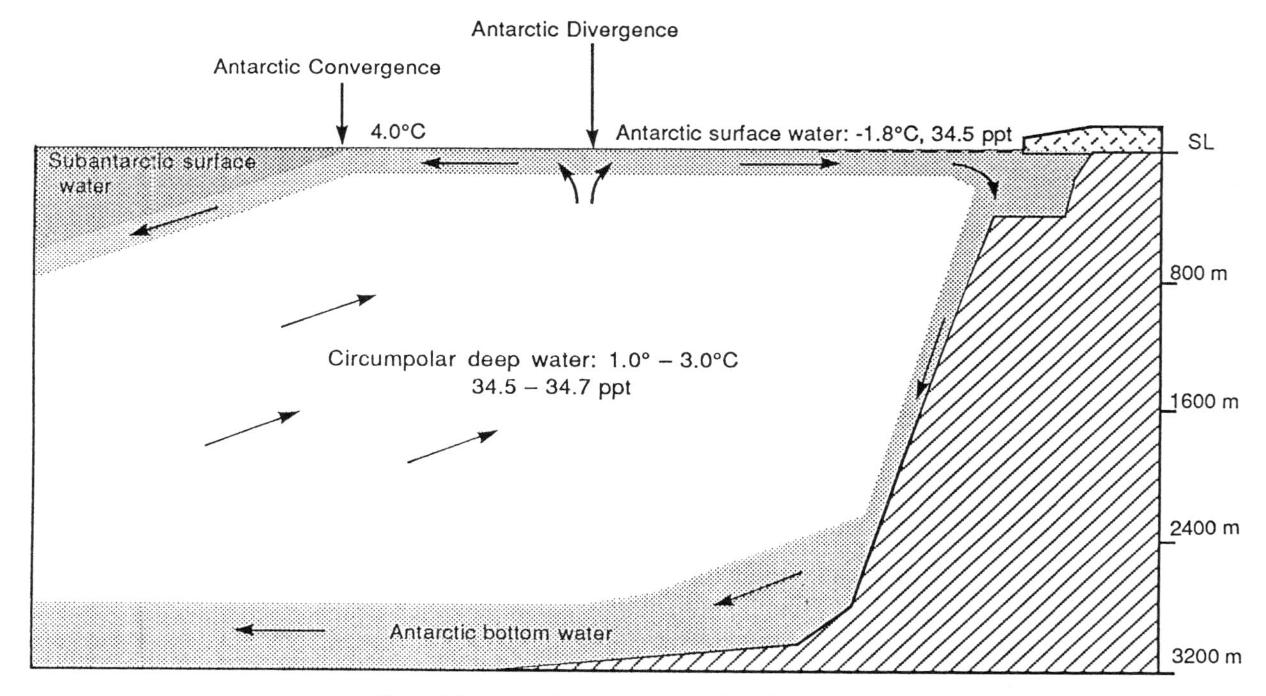

Figure 5.11 Antarctic water masses. After Hempel (1985).

narrow zone close to the continent where diverging westerly and easterly winds pull subsurface waters upward and spread them across the ocean surface. The water masses are illustrated in Figure 5.11.

The subsequent history of the upwelled water determines its main characteristics. That which spreads northward is warmed in summer to a maximum of about 4°C at the Convergence, where it sinks below still-warmer subantarctic surface water; beyond the boundary it continues to spread north as the subsurface Antarctic Intermediate Current. The southern component spreads over the continental shelf. In winter this water is chilled and frozen; ice formation releases cold brine which sinks and mixes with underlying water to form a distinctive water mass of high density and high salinity, called antarctic bottom water.

This creeps across the shelf and down the continental slope, spreading along the sea-bed; much of it is formed over the shelves of the Weddell and Ross bights, and from there it spreads far northward into the northern hemisphere. Between bottom and surface waters, forming the main mass of ocean, is a layer of warm deep water. Over 2000 m thick, warm deep water originates as surface waters of the temperate and tropical oceans, and flows southward beneath the subantarctic surface water. Emerging at the Antarctic Divergence, it replaces the water lost to the two northward-flowing currents. Characterized by temperatures of 1–3°C and salinities around 34.5–34.7 ppt, it provides a constantly-replaced, nutrient-rich environment for the plankton of surface waters.

Strong but shallow currents within the Antarctic Divergence carry ice and surface waters westward about the continent. Westerly-flowing currents entering the Weddell Sea cause the development of a strong gyre, which carries a stream of cold, ice-filled surface water clockwise and releases it northward past the tip of Antarctic Peninsula and the South Orkney Islands. This circulating cold water accounts for the anomalously low mean temperatures of eastern Antarctic Peninsula. In the Ross Sea similar forces develop a counter-clockwise circulation, causing the piling-up of pack ice off the northeastern corner of the bight and along the West Antarctica coast. The strong eastward currents generated by prevailing westerlies are most marked at the surface, but penetrate to impart an easterly flow to the full depth of the water column.

5.4.2 *Antarctic plankton*

Phytoplankton. In winter when the sun is low or absent, primary production at the ocean surface is minimal. Surface waters retain few algae from the summer crop; there is little light, and the amount present is insufficient to stimulate the cells to division. Waters that are ice-covered remain unstirred by wind, and all particles including algae tend to sink and not be replaced. The world's clearest sea water has been recorded in the Weddell Sea in early spring, clear enough for a Secchi disc to be seen at a depth of 79 m (Gieskes *et al.*, 1987; *Polar Record*, 1987). This far surpasses an earlier record of 53 m in the Mediterranean, and is close to the transparency of distilled water.

As the sun climbs higher in early spring, warmth and light return. More light is refracted below the surface, and algae, especially diatoms, rapidly become plentiful. Many of these first algae of the season are released from ice floes; as the pack ice melts, initially at its northern edge and later across its whole width, diatoms and other algal cells that have been trapped in its interstices are liberated into the sea and soon begin to divide rapidly. The surface waters, now well-stirred by wind, are high in nutrients, and primary productivity starts to rise. To the ice algae are now added open-water species, all proliferating; from September to January the amount of chlorophyll *a* present in surface waters climbs rapidly, keeping pace with incoming solar radiation (El-Sayed, 1971, 1985).

Over 100 species of diatoms are involved in this massive spring bloom, including both pennate (flag-shaped) and discoid forms, e.g. *Nitzschia, Thalassiosira, Thalassiothrix, Corethron, Odontella* and *Chaetoceros* spp. Present in lesser numbers are species of dinoflagellates, notably *Protoperidinium* and *Dinophysis*, and other algae including the silicoflagellate *Dictyocha speculum* and gelatinous *Phaeocystis poucheti.* In spring local accumulations of diatoms stain ice floes and the water immediately beneath them green or brown; quite commonly in summer the algae form dense soup-like concentrations extending over huge areas; El-Sayed (1971) reported a yellow bloom of *Thalassiosira tumida* covering 15 500 km^2 in the southwestern Weddell Sea. Many of the diatom species and even more of the dinoflagellates are endemic to antarctic waters.

Stocks and rates of production vary from one area of the Southern Ocean to another. Both tend to be high at the northern edge of the pack ice, over continental and island shelves where turbulence brings fresh resources to the surface, and at boundaries between water masses.

Both stocks and production rates rise to a peak in December toward the Antarctic Convergence, and in February or March in higher latitudes, for reasons that are far from clear. Sunlight is still plentiful, the waters continue to be well-mixed by turbulence, and nitrates and phosphates (which often limit aquatic production) remain abundant except in very local areas

(Hardy and Gunther, 1936; Balech and others, 1968). Fogg (1977) and others have suggested that other key elements or nutrients may be used up, but none has yet been identified as limiting. The grazing of increased stocks of zooplankton will by this time no doubt be making significant inroads (see below). In lower latitudes there is often a further slight rise in standing crop and productivity in autumn, followed by a rapid decline to midwinter.

Phytoplankton studies have generally involved nets with meshes greater than 35 μm, which are fine enough to catch the algae that predominate in most crops. For quantitative sampling, nets are often paired (Fig. 5.12) and fitted with meters that measure the flow of water through them. In high concentrations of algae the meshes become clogged, and their filtering efficiency is reduced; however, it is not usually difficult to estimate numbers of cells per cubic metre of sea water. Alternatively the standing crop present in a given mass of water (i.e. biomass per unit area or volume) can be estimated from the amount of chlorophyll (specifically, chlorophyll *a*) present (p. 112). To assume that all the chlorophyll is contained in diatoms and other algal cells of similar size may, however, be misleading. Recent studies using much fine filters, of mesh size 10–20 μm, indicate that a wide range of very small organisms, collectively called nanoplankton, are also present in surface waters. These are as yet little-known but clearly very important in the economy of the plankton. They may contribute as much as 70% of biomass and productivity locally (El-Sayed and Taguchi, 1981; Yamaguchi and Shibata 1982).

Even smaller particles of less than 2 μm, called picoplankton, may also play an important role in productivity of polar waters, and indeed throughout the world's oceans. For Antarctic waters El-Sayed and Weber (1986) reported that up to 70% of chlorophyll *a* in the water column was contained in particles smaller than 1 μm. In the view of El-Sayed (1987), one of Antarctica's leading marine biologists, 'the picoplankton are a very important but highly variable component of the Antarctic marine ecosystem'.

Zooplankton and pelagic fish. These are present in Antarctic surface and near-surface waters throughout the year, concentrating dramatically in the top 200 m in summer. Antarctic zooplankton is characterized by an almost complete absence of larvae of benthic animals, and surface layers are rich in individuals but poor in species; numbers of species tend to increase with depth (Knox, 1970). That the characteristic communities are widespread both latitudinally and longitudinally is not surprising.

Figure 5.12 Raising a bongo net—a double plankton net for quantitative sampling. Photo: B. Stonehouse.

Mackintosh (1960) draws attention to an important aspect of the annual vertical migration that many species follow; dispersing northward at the surface in summer, their descent to 400–600 m in winter ensures that they return south in the southward-flowing waters at those depths. There is also a constant westward drift in surface waters immediately surrounding the continent, and an eastward drift under the influence of westerly winds beyond the Antarctic Divergence, that spread species longitudinally and ensure circumpolar distributions.

The single most prominent invertebrate species is the shrimp-like crustacean *Euphausia superba.* Its popular name 'krill' originated with Norwegian whalers. Prevalence of krill in summer surface waters has given rise to a misconception that krill and other euphausiids constantly dominate the zooplankton. In fact copepods, notably *Calanoides acutus, Calanus propinquus* and *Rhincalanus gigas,* form the bulk of the zooplankton for much of the time (Voronina, 1966, 1968; Hopkins, 1971). The different life-stages of these species become prominent at different times of the year and at different depths; many species have marked diurnal (vertical) and annual (horizontal) migrations (Hardy and Gunther, 1936; Dinofrio, 1987). *Euphausia superba* is shown in Figure 5.12.

Medusae, ctenophores, chaetognaths (*Eukronia, Sagitta, Euchaeta* and other species), salps, pteropods and larval fish also are generally present, often locally in swarm concentrations.

Euphausiids, of which the largest are over 5 cm long, may form dense shoals several hectares in extent. Krill *E. superba* and closely-similar *E. frigida* are species prevalent across the central and northern zones of the Southern Ocean; smaller *E. crystallorophias* is most common in colder waters among the pack ice and immediately surrounding the Antarctic continent (Marr, 1962). Throughout summer in the richest areas of the Southern Ocean, notable north and east of Antarctic Peninsula, huge swarms averaging 40–60 m across are often seen. Though individually translucent and almost invisible, euphausiids have enough pigment and are present in sufficiently density to tint the water red or orange. Such swarms are fed on by mysticete (whalebone or baleen) whales, and experienced whale hunters sought the swarms of krill in the expectation that whales would be seeking them too.

Like the smaller copepods, euphausiids are mainly herbivorous, combing algal cells from the water and gathering them in a basket-like arrangement of bristles on their many-jointed forelimbs. They feed suspended in mid-water, but are versatile enough to browse also under the ice and on the sea-bed. Krill take algae in a wide range of sizes, probably from 2μm to 60μm or more, varying their diet by combing up small copepods as well (McClatchie and Boyd, 1983; Boyd *et al.*, 1984).

The life cycle of krill is adventurous, though incompletely known despite decades of intensive study; for summaries see Everson (1977); Miller and Hampton (1988). Breeding areas include the Bellingshausen Sea, Scotia Sea, and rich, disturbed waters around the island of the Scotia Arc. Krill eggs, 5–6 mm in diameter, are laid near the surface and sink to depths of over 1500 m; the first few larval stages (nauplii and metanauplii) develop at even greater depths before the young krill rise to feed close to the surface. Such important parameters as age at maturation and longevity are unknown, though it seems likely that krill breed two or three times in a lifespan of three to five years. Where they spend winter has hitherto been uncertain; there is now strong evidence that most of the stock moves south in autumn and winter immediately under pack ice (Marschall 1988), where they browse on algae.

Other prominent omnivores and carnivores in the zooplankton include the larger copepods, chaetognaths, salps and larval fish. Less prominent, but quite possibly no less significant in energy pathways, are masses of heterotrophic choanoflagellates, bacteria and other small-to-microbial organisms, the presence of which, like that of the very small autotrophs, has only recently been perceived (El-Sayed, 1987).

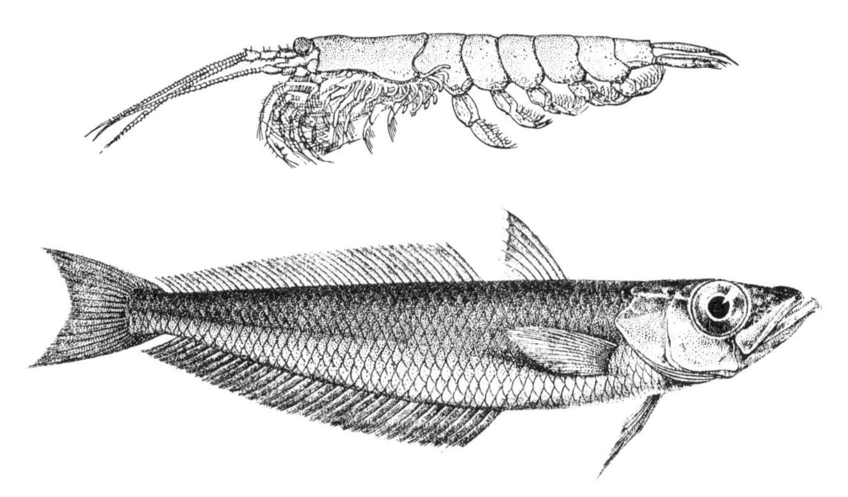

Figure 5.13 Key pelagic species in Antarctic marine food webs. Upper: krill *Euphausia superba*, one of a genus of antarctic shoaling crustaceans; length 4 cm (after Miller and Hampton 1988). Lower: *Pleuragramma antarcticum*, a shoaling fish; length 25 cm (Boulenger, 1902). Several species of seals and seabirds depend on these for summer feeding.

Among surface of predators are the pelagic shoaling fish *Pleurogramma antarcticum* (Figure 5.13), which forms an important link between the plankton and the higher vertebrates.

5.4.4 *Mid-water, sea-bed and shore*

Mid-waters of the Southern Ocean are little explored; for long considered empty, they are now known to be the haunt, possibly the refuge, of several species of fish and squid which feed either near the surface or at the bottom. Many are fast-moving and able to avoid trawl nets, but less adept at avoiding penguins and seals; most of the smaller species are better known from the stomach contents of predators than as independent catches.

Shores of the Antarctic continent and the maritime region, like those of the Arctic, are subject to ice scouring, which neither plants nor sessile animals are able to survive. Where islands beyond the northern limit of pack ice (for example South Georgia) have a rich intertidal flora and fauna, and are ringed by beds of giant kelp *Macrocystis gigantea* with strands 20–30 m long, those within the pack ice zone have a relatively bare shoreline and restricted sublittoral flora. Anchor ice (see above) grows freely from the sea-bed in high latitudes.

Studying the littoral zone near Palmer Station, on the South Shetland Islands, Hedgpeth (1971) noted considerable seasonal variation. Granite rocks that were almost bare in early spring had by late summer acquired a thin growth of brown diatoms (*Navicula, Fragilaria, Achnanthes, Licmophora*) and filamentous green algae (*Ulothrix, Enteromorpha, Cladophora, Monostroma*), growing upward in a narrow band upward from mean low tide level. Below them, and clearly marking the low-tide level, extended a perennial covering of pink crustose algae (*Lithothamnion, Lithophyllum*). Among the green algae were small crustaceans and worms, and in a narrow zone above were small specimens of red algae (*Curdiea, Iridae, Leptosomia*), growing in crevices during the summer only. A single browser, the limpet *Patinigera polaris*, lived mainly below low-tide mark but emerged to scour the whole zone; those close to sea level were predated by dominican gulls *Larus dominicanus*.

From the sublittoral zone downward the sea-bed is covered by fine debris. The continental shelf and deeper sea-bed beyond, to a distance of 300–600 km from the shore, have a covering of fine grey glacial mud, forming a layer that varies in thickness from a few metres over active areas of sea-floor spreading to 200 km or more in older, more stable regions. Within the mud is a scattering of angular boulders—rock fragments that have been carried from the land by icebergs and dropped as the bergs weather and melt. Further north the mud is siliceous, enriched by a constant rain of diatom casings from the phytoplankton.

In waters north of the Antarctic Convergence the bottom mud becomes calcareous, due to the rain of foraminiferan casings from the surface. Cores drilled from these muds provide an excellent long-term record of changes in the distribution and species composition of algae and foraminifera, from which changes in surface water temperatures may be deduced.

These thick accumulations of mud provide substrate for a rich sea-bed fauna, predominantly of water-filterers (sponges, molluscs, salps) and scavengers (worms, arthropods, pycnogonids, fish).

5.4.5 *Birds and mammals*

The Southern Ocean is the feeding ground of many thousands of seabirds, seals and whales, some present only in summer, others throughout the year.

Seabirds. Of 38 species of birds that breed south of the Antarctic Convergence, seven are penguins, four albatrosses, 20 smaller petrels, and seven others include gulls, skuas, terns and cormorants. Of these, 10 species breed on the Antarctic continent and a further five on Antarctic Peninsula; the rest breed on the islands and archipelagos both north and south of the limit of pack ice. Within the Convergence only five species of birds feed on land or in fresh water, all on islands close to the Convergence; they include two species of pintail ducks, two of sheathbills and a single species of pipit (Stonehouse, 1972). Details of breeding, feeding ranges and other data on all antarctic and subantarctic species are summarized in Watson (1975); see also Stonehouse (1985b), Siegfried (1985) and Croxall (1984).

On continental Antarctica two species of penguins, emperors *Aptenodytes forsteri* and Adélies *Pygoscelis adeliae*, are the most prominent breeding birds; both breed colonially in groups that may number tens or hundreds of thousands. Emperors are winter breeders, gathering on inshore sea ice to lay their single eggs in May and June, incubating and brooding chicks on their feet through the coldest months in temperatures that may drop below – 50°C. Adélies breed ashore, mostly on raised beaches close to sea level from November onward; they lay two eggs in pebble nests in October and November, rearing their chick through summer. McCormick's skuas *Catharacta maccormicki*, brown predatory gulls (Figure 5.14), breed among the Adélies in summer, feeding partly at sea but also scavenging on the penguin colonies for abandoned eggs and chicks. Several species of petrels including all-white snow petrels *Pagadroma nivea*, mottled pintado petrels *Daption capensis* and tiny Wilson's storm petrels *Oceanites*

Figure 5.14 McCormick skuas—Antarctic scavengers and predators. Closely-related species occupy similar niches in the subantarctic and in the Arctic. Photo: Guy Mannering.

oceanica breed in cavities on scree slopes and under rocks. Many breed on coastal cliffs, but small colonies are reported on isolated nunataks tens or even hundreds of kilometres from the sea.

The maritime antarctic region has Adélie and closely-related gentoo and chinstrap penguins *P. papua* and *P. antarctica*, a second species of skuas *C. lonnbergii*, dominican gulls *Larus dominicanus*, blue-eyed cormorants *Phalacrocorax atriceps*, and a wider selection of petrels that breed both at the surface and in cavities among scree. Periantarctic islands lying north of the pack ice zone have the broadest range of breeding seabirds, including king penguins *A. patagonica*, crested penguins *Eudyptes chrysolophus* and *E. crestatus*, wandering albatrosses *Diomedea exulans*, two or three species of mollymauks (smaller albatrosses), and several species of burrowing petrels (for example prions *Pachyptila* sp., grey petrels *Adamastor cinereus* and white-chinned petrels *Procellaria aequinoctalis*) that nest deep in the peaty soils.

Most of these seabirds appear on land only during spring and summer, breeding when surface waters are richest in plankton; for the rest of the year they disperse widely. Adélie penguins winter in the pack ice; most others fly to open water north of the ice, where they circle the west wind belt. Brown and McCormick's skuas and Wilson's petrels fly as far as temperate latitudes of the northern hemisphere. During the summer breeding gulls, skuas, terns and cormorants feed close to land; the rest range widely over the ocean, browsing at or close to the surface in single-species or mixed flocks. Though all are feeding from the same source, and to some degree feed on whatever is most plentiful, competition is minimized by selective feeding; Croxall and Lishman (1987) and Prince and Morgan (1987) summarize methods of feeding respectively among penguins and petrels in southern waters.

Antarctic seals and whales. The six species of antarctic seals are summarized in Table 5.2; for surveys of their ecology see King (1983), Laws (1984) and Bonner (1985). There are no southern walruses. The largest antarctic seals are southern elephant seals, which breed on cool temperate and antarctic fringe islands; immature animals appear on antarctic mainland coasts in summer. These are probably the deepest divers, feeding mainly on fish and squid. Intensively hunted during the nineteenth century, they have more recently been managed and subject to controlled hunting on South Georgia (Laws, 1960). Ross and crabeater seals live solitary or in widely-dispersed communities on the pack ice, usually toward the outer edge, feeding largely on krill. With a total population estimated at 15–30 million, crabeaters are almost certainly the world's most numerous seals (Laws, 1984); their wide dispersal on pack ice protects them from human depredations.

Leopard seals, which feed on krill, fish, squid, young seals of other species (notably crabeaters), and penguins and other birds, also occupy the outer edge of the pack ice, but range widely from the cool temperate islands to Antarctica. Weddell seals, which feed mainly on fish and squid, breed mainly on fast ice close to the shore of Antarctica and fringe islands, and spread to the pack ice in late summer. Breeding groups of over 2000 have recently been found among the shelf ice of the Weddell Sea (Hempel and Stonehouse, 1987); a small breeding population occurs on South Georgia. Southern species of fur seals are island breeders (Figure 5.15). Formerly plentiful on antarctic fringe and peripheral southern islands, they were hunted almost to extinction by commercial sealers during the late eighteenth and nineteenth centuries. Stocks have recently recovered; populations are expanding on South Georgia and many of the Scotia Arc islands (Bonner, 1985) and a newly-established breeding colony has

Figure 5.15 Part of a breeding colony of Antarctic fur seals, Bouvetøya. Photo: B. Stonehouse.

recently been reported on Iles Kerguelen (Jouventin and Stonehouse, 1985).

Several species of dolphins and small whales, including killer whales, Commerson's dolphins *Cephalorhynchus commersonii*, dusky dolphins *Lagenorhynchus obscurus*, hourglass dolphins *Lagenorhynchus cruciger*, and southern beaked whales *Berardius arnuxii* have been recorded in the Southern Ocean, mainly in ice-free waters. Among the larger whales, southern stocks of sperm whales (males only), humpbacks and four species of rorquals—blue, fin, sei and minke whales—are regular summer visitors to the rich waters north of the receding pack ice (Gambell, 1985). They penetrate south to varying degrees, blue and minke whales often appearing well south of the pack ice edge. Minkes, together with southern bottlenose whales *Hyperoodan planifrons* (Figure 5.16) and killer whales have been reported making use of late winter polynyas several hundred kilometres within the pack ice edge (Hempel and Stonehouse, 1987).

Humpbacks, sperms and rorquals formed the basis of an extensive antarctic whaling industry, which began in 1904 at a single whaling station on South Georgia and expanded enormously over the next three decades,

Figure 5.16 Southern bottlenose whale, Weddell Sea, Antarctica; an almost identical form inhabits north polar seas. Photo: B. Stonehouse.

making use of both shore stations and pelagic fleets. Though controlled in part by international agreement, the industry overfished its stocks and declined during the 1950s and 1960s. For a brief history see Gambell (1985); for fuller coverage see Tønnessen and Johnsen (1982). The removal of many thousands of large whales, especially the krill-feeding rorquals, from the Southern Ocean within a few decades is thought by some ecologists to have favoured the many other species, including fish, birds and seals, that feed on krill.

5.5 Summary and conclusions

Polar seas, dominated by seasonal sea ice, are ecologically poor in winter and very much richer in summer, especially around their fringes and where disturbance or upwelling enhance nutrient supplies. The structure, water masses and sea ice cover of Arctic and Southern Oceans are compared; the central Arctic Ocean, permanently ice-covered, has no direct equivalent in the south, except in the much smaller gyres of the Weddell and Ross Seas where pack ice circulates. Despite their differences, the two oceans support basically similar food webs, with high seasonal productivity from small overwintering standing crops. The possible significance of marine microflora, hitherto neglected, is discussed. Deep-water and mid-water communities are surveyed briefly, including the little-known but important cephalopods. Prominent at both ends of the earth are large stocks of fish, seabirds, seals and whales, which maintain indigenous human communities and attract commercial enterprise.

CHAPTER SIX
ACCLIMATION, ADAPTATION AND SURVIVAL

6.1 Organisms and polar environments

Though the earth has always had poles of rotation, the polar environment has not always been cold (Chapter 1). High latitudes were temperate until cooling began in the mid-to-late Pliocene; overall cooling from then to the present has been punctuated by spells of warming. As isotherms swung north and south during repeated cycles of warming and cooling, the temperate biota bordering the cold regions migrated latitudinally with them. Intensification of cold allowed the arctic flora and fauna to spread southward; amelioration encouraged repossession of the borderlands by temperate species.

Natural selection for polar flora and fauna has never, therefore, been simply the selection of cold-hardy species for increasingly cold environments. Many other factors not directly related to cold have been involved. The wobbling but persistent angle between earth's axis and ecliptic (Figure 1.2) has ensured that, though not always icy, polar regions have always been strongly seasonal. Adaptation to thrive in seasonal environments, in which day-length, temperatures, water and food availability and other important factors fluctuate widely throughout an annual cycle, may be at least as critical as adaptation to cold, and more difficult for species to adjust to by natural selection. The ability to fit life-cycles into seasons, to flourish in times of plenty and survive in times of dearth, is clearly critical for polar species; so also are such qualities as adaptability to change, avoidance of specialization, and deftness in colonizing new habitats.

6.1.1 *Acclimation and adaptation*

These two important adjustive processes occur simultaneously but within different time-scales. Acclimation is the non-heritable modification of characters caused by exposure of organisms to environmental changes. It may occur several times within an organism's lifetime, sometimes within a span of hours or even minutes. Individuals acclimatize to cold, for example, by adjusting physically, physiologically or psychologically following cold exposure. Such changes are ephemeral, readily reversible and not known to effect the genome.

Adaptation has two linked meanings, both implying longer-term hereditary changes. An adaptation is a modification of the genome resulting in structures, functions or behaviour patterns that increase the probability of an organism surviving and reproducing in a particular environment. Thus, all warm-blooded animals are insulated, but the special insulation of polar birds and mammals, additional to that of their temperate or tropical counterparts, is an example of adaptation in this sense. The time-scale involved is seldom measurable, but spans many generations. Adaptation in the second sense is the accumulation, in a species, of enough of these changes to promote viability within the specified environment. Polar plants and animals are judged to have acquired adaptation—to have adapted—when they are demonstrably successful within their environment.

Though individual acclimations are not inheritable, ability to acclimatize may well be inherited. Plants and animals that were incapable of acclimatizing to cold would not have survived in the cooling environment of the early Pleistocene. Those that were able to acclimatize might have survived long enough for hereditary processes to be invoked and adaptation to occur. Their facility in acclimatizing may still be called into frequent use during the lifetime of individuals, for example in the annual frost-hardening of herbaceous plants.

6.1.2 *Polar adaptation*

Many species that thrive in polar habitats are found also in subpolar and temperate habitats. When polar stocks of such cosmopolitan species are recognized as differing in form or physiology from temperate stocks, they are often assigned to separate subspecies or races. Polar stocks are generally assumed to have accumulated, and still be accumulating, adaptations that equip them for the polar environment; endemic polar species, that are found *only* in polar regions, are assumed to be derived from temperate or subpolar stocks, via intermediate stages that no longer exist. Many polar species have clearly originated in this way by adaptations of metabolism, form and lifestyle. Their penalty for doing so is frequent failure, for polar habitats are harsh and highly variable, unbuffered and unpredictable. The rewards for the enterprising are abundant, though seasonal, energy for the asking, in relative freedom from competition.

Polar adaptations are usually developments or extensions of adaptations that organisms have acquired elsewhere to meet other challenges; only

Figure 6.1 Cormorants, which dive to catch fish and benthic invertebrates, are widely distributed in temperate waters. Well insulated by fat against steep temperature gradients, they are pre-adapted for life in polar waters. Photo: British Antarctic Survey.

secondarily and relatively recently have they acquired polar relevance. Lichens that survive cold and desiccation on mountain tops in temperate regions may find tundra winters only marginally more difficult, and tundra summers an unaccustomed blessing; the adaptations they already possess for harsh temperate environments become pre-adaptations for polar living. Aquatic birds that spend much of their time in water, well insulated by fat and waterproof plumage (Figure 6.1), are pre-adapted for life in cold lands; the insulation of marine mammals in temperate seas pre-adapts them for polar seas. It is not surprising to find lichens, penguins, seals and whales well represented in polar regions, and varying only slightly, if at all, from non-polar congeners.

The adaptations of form, physiology and behaviour of polar species, whether acquired early or late in their evolution, are combined in solving the problems presented by the polar environment. Strategies for survival adopted by some particular groups of organisms are summarized below.

6.2 Survival in plants

6.2.1 *Resistance to freezing*

Conditions of temperature and light in polar regions differ in amount but not in kind from those of temperate regions. Polar plants must be capable of being frozen for several months each year, and to undergo repeated freeze–thaw cycles throughout the growing season, even at the height of summer. They must be able to cope with aridity, high daily and seasonal temperature ranges, seasonal swings in photoperiod from long days to long nights, and a growing season ranging in length from three to four months on the polar fringes to a few days or even hours in higher latitudes. These problems, and some of the strategies adopted by plants, for dealing with them, are discussed fully by Larsen (1964), Bliss (1971), Savile (1972), Courtin and Mayo (1975) and Heide (1984).

Frost-hardening—the process that enables plants to withstand freezing conditions without damage—is a property of cells that develops seasonally in response to climatic factors. Its mechanisms have been studied most fully in vascular plants of commercial importance, for example cereals, green vegetables and shelter-belt trees, in temperate boreal regions that are subject to hard winter frosts (Kaurin *et al.*, 1985). Similar mechanisms are likely to be involved in all other multicellular plants in which hardening occurs.

Most tropical plants have no resistance to freezing and are killed by even slight frost. Cell sap normally starts to freeze at about −3°C, and ice forming within the cells causes fatal damage to plant tissues. Most temperate species tolerate subzero temperatures to one degree or another, initially by avoiding tissue freezing but also by being able to tolerate and contain ice in their tissues. Avoidance involves allowing cell sap to concentrate, so reducing its freezing point; in this way frost-hardened plants may remain ice-free and physiologically active for many days at temperatures well below 0°C. Frost-hardening may affect some tissues but not others; it is not unusual for well-established tissues to be hardened but for buds and growing shoots to be nipped by frost.

Frost-hardening is an acclimation process invoked by environmental stimuli. The stimuli vary with species but include cooling and decreasing daylength: wavelength and intensity of light are also implicated (Kaurin, 1984). Tronsmo and Kaurin (1985), present a simplified model of the process. Cells that have a genetic capacity for hardening are stimulated by the environment to produce isoenzymes—alternatives to those normally in production—that change the phospholipid and sterol content of the cell membranes. The isoenzymes, which may be derived from alternative gene sets provided by polyploidy, in effect increase membrane permeability and allow water to pass out of the cells, concentrating cell solutions and lowering their freezing points. Simultaneous changes in the cell walls, hormonally induced, produce negative pressure potentials during extracellular freezing, which control the dehydration. Several other factors affecting protein bonding within the plasma membranes may be involved (Levitt, 1962; Siminovitch, 1969).

By concentration of cell contents and supercooling, frost-hardened plant tissues readily survive temperatures down to −10°C. Freezing is further inhibited when the cell sap is divided into several vacuoles rather than a single large one. Ice begins to form due to the presence of bacteria or other nucleators in the cell sap. When freezing occurs, over 60% of the freezable water becomes frozen at −4°C, practically all of it at −15°C. By further concentration, and by various methods of isolating the cooled, concentrated cell sap from contact with nucleators, some plants can withstand supercooling in at least some of their tissues to temperatures as low as −55°C.

Many hardy herbaceous and woody plants tolerate ice in their tissues, but are killed by build-up of ice due to increasing cold or length of frost. Gusta (1985) has reviewed the sequence of events. When a plant starts to freeze, ice accumulates first in interstitial water, leaf veins and other extracellular spaces. The presence of ice withdraws water in vapour form from distal cells, concentrating their solutions and building up the extracellular crystals. The dehydrated plasma membranes adhere to the cell walls, causing them to collapse; growing ice crystals rupture the cell walls and membranes and destroy cytoplasmic integrity. On thawing, live cells rehydrate and return to their original shape; damaged ones fail to recover and die. Physiological and biochemical aspects of the process are reviewed in more detail by Kacperska (1984).

The converse process of de-hardening, induced by exposure to warmth, is less clearly understood, but is probably not a simple reversal of hardening. Temperate species are slow to accept it; those that de-hardened and resumed growth at the first sign of spring warming would suffer severe damage in subsequent frosts.

Temperate species of vascular plants that invade the low Arctic are often damaged or killed by frosts in summer; those native to the high Arctic withstand repeated severe summer frosts without damage. From the Canadian Arctic Porsild (1951) records catkins forming on *Salix richardsonii*, one of the dwarf willows, during an unseasonable warm spell in April; the catkins were frozen solid during a subsequent three-week spell of intense cold, but resumed flowering immediately afterwards and set seed in June. Savile (1972) comments that arctic plants may be brittle with frost, sheathed in ice or buried in snow during the summer period of most rapid growth, and yet able to resume growth immediately and without injury as soon as warm weather returns. Though obviously similar, summer hardiness appears to be a more rapid and flexible process than the hardening and de-hardening of autumn and spring; it is an adaptation that every species of the high polar region is likely to need.

6.2.2 *Non-vascular plants*

Algae, mosses, liverworts and lichens have many preadaptations but few special adaptations for polar life. In the Arctic, few northern polar taxa are sufficiently differentiated to be called endemic species. For example, of 134 species of mosses identified in Peary Land, north Greenland, Holmen (1960) categorized only one, *Ortothecium acuminatum*, as endemic to the area, and only 24 (18%) as having a mainly arctic distribution; the remaining species are widely distributed in the north temperate zone. Southern hemisphere cryptogams are relatively poorly collected, with confused phytogeographic relationships (Smith, 1984; Longton, 1985). Among 105 species of maritime and continental antarctic mosses (Smith, 1984) only a few are currently regarded as endemic, and the number declines as 'antarctic' species are discovered in South America and elsewhere. Most of those collected on the continent are well represented in temperate southern latitudes; some are cosmopolitan, and a few ecologically prominent species are bipolar (Longton 1985).

Similar relationships are found among polar lichens and algae, suggesting that temperate latitudes provide a reservoir of pre-adapted non-vascular plants, from which polar regions are constantly colonized and recolonized. The single most important pre-adaptation common to these forms is their ability to cope with drought by partial dehydration, and to resume active metabolism quickly on rehydration. This is a useful quality in any dry climate, warm or cold. By conferring additional resistance to cellular frost damage, it is especially useful in alpine, steppe and boreal forest regions, and no less so for polar conditions. The non-vascular cryptogams have wide climatic tolerance, and spread readily by wind-distributed spores or vegetative propagules. Most plant bodies are small and compact, allowing full advantage of warmth at the boundary layer between rock and atmosphere, and protection against wind erosion under a thin snow-covering.

Adaptation in bryophytes has been summarized by Savile (1972). Many species are genetically variable and thus readily adaptable; many are monoecious and self-fertile, and able to penetrate new areas without completely sacrificing genetic recombination. Though not aggressive colonists, they reproduce freely by vegetative growth and thrive in a range of conditions that other plants find intolerable. Antarctic mosses have survived three years in darkness at $-15°C$ (conditions that might occur under long-lasting snow-drifts) and grow readily in the miniature greenhouses that form under thin snow or translucent ice.

In moss turf (*Polytrichum alpestre*) and moss carpet (*Drepanocladus uncinatus*) communities on Signy Island, South Orkney, Collins (1977) found growth starting before snow-melt, though most occured during the four snow-free months between November and March. Polar plants grown experimentally in temperate regimes thrive better and produce more per unit area, performing best in temperatures 5–20°C higher than at home. Mosses growing on Signy Island in summer at temperatures between $-5°C$ and $+5°C$ have an optimal net photosynthetic rate at 10–15°C; the rate is depressed at temperatures above 15°C, which may be experienced across midday in high summer. Net annual production of 400–99 $g\,m^{-2}$ dry weight is reported from Signy Island, values comparable with those recorded in both temperate and Arctic tundra communities.

Similarly, in the subarctic moss *Dicranum elongatum* Heino and Karunen (1985) found a temperature optimum for photosynthesis of about 10°C, though there was little variation within the range 5–15°C. This species photosynthesized actively throughout the year on a mountain slope in Finnish Lapland (69° 45′N). Production was highest in August, decreasing in late autumn to low winter and spring levels; rates were quoted as ‘quite high’ even in temperatures close to freezing point.

Lichens, generally prevalent where soils are poor or absent, grow in conditions of aridity, exposure, and abrasion by wind-blown sand and snow that no other plants tolerate (Lamb, 1970; Dodge, 1973). Some 60 out of 90 antarctic genera are crustose, producing a flat thallus closely appressed to the substrate; only 17 genera are foliose (i.e. with a divided, leaf-like thallus), and only 15 genera, occupying the least exposed positions, are fructicose—short, cylindrical, erect or recumbent, and attached only at the base (Dodge, 1973).

Growth, standing crop and metabolic rates of lichens are difficult to measure, but these unusually tough plants seem particularly well pre-adapted to polar conditions. They respire at lower levels of light, temperature and tissue moisture than bryophytes or vascular plants, and are protected by dehydration against exhaustion of carbohydrate reserves (Smith, 1984). Photosynthesis occurs at temperatures as low as −10°C and very low ambient light levels (Gannutz, 1969, 1970). Extreme cold is no additional burden; lichens have survived after experimental plunging into liquid nitrogen at −196°C (Kappen and Lange, 1972). Optimal temperatures for photosynthesis measured in polar species are generally lower than those quoted for mosses and closer to ambient temperatures (Lange and Kappen, 1972; Kappen *et al.*, 1981; Kappen and Friedmann, 1983). Smith (1984) tabulates productivities in fructicose lichens of up to 1130 g m^{-2} dry weight in South Georgia, and up to 250 g m^{-2} on Signy Island.

6.2.3 *Vascular plants*

The arctic vascular flora includes some 1899 species and subspecies, representing 404 genera and 106 families (Löve and Löve, 1975). Non-flowering plants include 61 taxa (species and subspecies) of pteridophytes and 10 of gymnosperms; flowering plants include 511 taxa of monocotyledons and 1317 of dicotyledons. Those of the low Arctic (within a boundary similar to that of Aleksandrova, discussed on p. 85) appear mainly to be boreal alpine species that have spread northward. Some spread after the last glaciations, others have been there longer after surviving the glacial periods in ice-free refugia (Lindsey, 1981).

Vascular plants of the high Arctic seem by contrast to be remnants of the Tertiary flora that have survived in refugia throught all the glaciations and recently dispersed southward, though seldom beyond the low Arctic zone. Some 24% of them are true endemics; 18 species of flowering plants in this category are restricted entirely to the high Arctic. For a discussion of vascular plant distributions within the Arctic see Haber (1986).

Many arctic vascular plants are polyploid; multiple chromosomes appear in almost 60% of low arctic species and 70% of high arctic species (up to 80% of high arctic endemics). Polyploidy may enhance adaptability, and is regarded as useful preadaptation for stressful environments (Löve and Löve, 1974).

Plant form. Though widely distributed across the Arctic, and included among the northernmost of all vegetation, vascular plants are readily susceptible to damage by wind and frost. Savile (1972) draws attention to the role of strong winds and abrasion in determining their distribution and habits. Like mosses and lichens, the vascular plants adopt low profiles, colonizing sheltered sites away from prevailing winds, where they are protected by snow in winter. Herbaceous forms grow as tightly-packed mats (*Saxifraga, Stellaria*), rosettes (*Draba, Potentilla*) or cushions (*Silene*), habits that keep them under the snow in winter and close to the warm soil surface in summer. Winter buds, containing the vulnerable growing points, are seldom covered by protective bud scales, which might impede early spring growth. More often they rely on whorls of old leaves, tightly-packed neighbouring shoots, and above all snow-cover, to protect them from wind abrasion. Monocotyledons, including grasses and sedges, form tussocks of hard, often silicious leaves that protect the growing points within.

Woody shrubs too depend on winter snow for protection against extreme cold and wind abrasion, which shape them by killing exposed growing points. While much of Siberia, Alaska and the western Canadian Arctic have well-developed shrub tundra, strong winds and low snowfall keep much of the central and eastern Canadian tundra almost free of emergent shrubs (Savile, 1968). At Chesterfield Inlet, northern Canada, in these conditions, Savile and Calder (1952) found willow shrubs (*Salix* spp.) and dwarf birch *Betula glandulosa* growing only 15 cm high on open ground; in the shelter of boulders the willows grew to 1 m or more, but in exposed areas they became completely prostrate, with horizonal laterals close to the ground. They thus became barely distinguishable from the genetically prostrate species of willow, for example *S. arctica*, which grows to the extreme northern limit of land (Savile, 1972).

Rosettes, mats, cushions and tussocks create internal microenvironments that are more favourable to growth than the world outside; temperatures and humidities measured among growing shoots in spring are usually higher than ambient, though soil and root temperatures may remain lower. Dark pigments in leaves and flowers contribute to warming; their sparing use in spring leaves suggests that green pigments alone are sufficient to bring the leaves to optimum temperatures. Mountain sorrel *Oxyria digyna* and tufted saxifrage *Saxifraga caespitosa* typify highly variable species; the colder and more exposed the site, the darker, more heavily pigmented are the stems and leaves (Savile, 1972). The pigmented, parabolic flowers of arctic poppies *Papaver* spp., buttercups *Ranunculus* spp. and several other genera follow the sun, focusing its rays at their centre; this provides a warm spot attractive to pollinating insects, and raises the temperature of maturing anthers and ovaries.

Pubescence (hairiness) characterizes many alpine species (Hedberg, 1964); stem and leaf hairs reduce air circulation and water loss, protect leaves from over-bright sunlight and trap outgoing heat. Only a few arctic species (perhaps of recent alpine derivation; Savile 1972) are densely hairy, but many use hairs strategically to good advantage. The almost transparent hairs on catkins of *S. arctica* and other willows, for example, coupled with dark pigmentation, provide greenhouse conditions that help to warm the underlying tissues and enhance seed production (Krog, 1955).

Growth. Growth is generally slow; seasonal growth in widely-distributed species is faster where the climate is warmer. *Salix arctica*, for example, grows three times faster in East Greenland than at Resolute, Cornwallis Island, and barely survives at Isachson, Ellef Ringnes Island; July mean temperatures are respectively about 8°C, 4°C and 2°C (Warren Wilson, 1964). Arctic stocks of *Oxyria digyna* (Mooney and Billings, 1961) and meadow rue *Thalictrum alpinum* (Mooney and Johnson, 1965) have lower optimal photosynthetic temperatures than alpine stocks of the same species. Some arctic species appear to be markedly temperature-sensitive in rate of growth, others less so, though low temperatures generally reduce the total biomass attained by cold-tolerant species in a single season (Savile, 1972). Limits may be set also by deficiencies of nitrogen and other key minerals in the soils. For methods of measuring photosynthesis, gas exchange, growth and production in high arctic vascular plant communities see Mayo *et al.*, (1977), Addison (1977) and Whitfield and Goodwin (1977).

Dwarfism, both genetic and phenotypic, is marked among high-latitude flowering plants. Where the growing season in short, a successful strategem of many plants is to remain small, grow short flowering stems, and invest their small savings of energy in a few well-endowed seeds. The short season available for growth restricts annuals and biennials mainly to the arctic fringe; one notable exception, *Koenigia islandica*, produces tiny plants 4–5 cm high in especially favoured areas as far north as Devon Island and northern Greenland, where they set seed annually. Perennials succeed by responding early to spring. Few are fully evergreen; rather more keep a small proportion of semi-evergreen leaves (usually the last to be produced in summer; Savile, 1972) that allow them to start photosynthesis as soon as the sun returns. Many store food overwinter in fleshy stems, roots and leaf-bases; deep storage roots are restricted to dry areas of the subarctic, where the spring thaw comes early.

Most perennials grow periodically, stopping before the end of the growing season when they have reached a particular stage, for example the production of winter buds. A minority, including many grasses, grow aperiodically, continuing until stopped only by bad weather in autumn and resuming in spring. Periodic growth ensures that plants reach safe overwintering stages, and do not risk the loss of tender new tissues to frost. Aperiodic growth takes full advantage of long seasons, but risks the loss of unseasonable shoots or flowers. The distinction is blurred by species that grow periodically in hard conditions and aperiodically where they are protected. The crucifer *Braya humilis* was originally described as aperiodic in northeast Greenland (Sørensen, 1941), where flowers produced in late autumn were retained frozen but viable in snow until the following spring; the same species was found to be periodic, with a late summer growth check, in the harsher snow-free conditions of Hazen, north Ellesmere Island (Savile, 1972).

Growth and maturation of vascular plants are often controlled by light, usually in conjunction with temperature. Dormancy is a response to short day-length and low temperatures; growth begins as days lengthen and temperatures increase. Spring bud-break, flowering, the formation of next year's buds, germination, stem growth, and the development and movements of leaves all depend on spectral quality or quantity of ambient light. Polar plants appear to have adapted positively to the long days, long nights, low light intensities and other special conditions (Chapter 2) of their environment. Experiments show that many tolerate a wide range of light conditions, but a few high polar species perform best in long day regimes, or refuse to flower in anything less than a 22–24-hour day. We do not know how many species of temperate latitudes have been unable to adjust, and so disqualified themselves as polar colonists, or to what extent existing polar plants are restricted in their distribution by light. For a review of plant adaptations to light environments see Salisbury (1985).

Reproduction. Arctic flowering plants are mostly capable of cross-fertilization, and make use of both wind and insects in pollination. Many are also self-fertile, or form seed by agamospermy, in which diploid embryos appear in seeds without benefit of cross-fertilization. In one or other of these ways, seeds are usually produced abundantly in normal seasons. Most seeds are spread by wind; a few rely on birds or mammals for dispersion. Rhizomes, runners and bulbils are alternative forms of vegetative reproduction common in grasses and sedges, and known also in many dicotyledons. These methods allow plants to spread in marginal regions where seeding is uncertain. Like agamospermy, they yield offspring that are genetically identical with the parents. Inbreeding appears to outweigh crossbreeding in many stocks of arctic plants, a condition that may be adaptive in a harsh environment where interspecific competition is insignificant. For discussion of the genetic implications of this interesting possibility see Savile (1960, 1972) and Mosquin (1966).

The two species of antarctic vascular plants, antarctic hair grass *Deschampsia antarctica* (Figure 6.2) and antarctic pink *Colobanthus quitensis*, occur patchily on Antarctic Peninsula south to the Terra Firma Islands, and on the South Orkney and South Shetland Islands. Only *D. antarctica* has so far been found on the South Sandwich Islands (Smith, 1984; Smith and Poncet, 1985), but both occur on South Georgia. In the maritime Antarctic they form short mats or cushions up to 25 cm across and 5–10 cm deep, typically on warm north- or west-facing slopes close to sea level, in association with each other and with mosses; they are usually underlain with brown soils to which they have contributed roots and other organic material (Longton, 1985). On South Georgia the hair grass forms extensive meadows, the pink grows in fellfields associated with

Figure 6.2 Part of the most extensive sward of Antarctic hair grass (*Deschampia antarctica*) in the Antarctic. Lynch Is (SPA No 13), S. Orkney Is. Much of this area was destroyed in 1987–88 by fur seals, which are increasing in numbers. Photo: Ron Lewis Smith

shrubs, grasses and mosses (Bonner, 1985). For studies of the distribution and physiology of these species at various localities throughout their range see Edwards (1972, 1973, 1974, 1975); for a study of experiments in introducing other vascular plants from South Georgia to the maritime Antarctic see Edwards (1980).

6.3 Survival in invertebrates

The microfaunae of polar soils and freshwater environments include representatives of many invertebrate phyla. Soil fauna studies in groups other than mites and insects have generally been restricted to systematic collection, identification, taxonomy, basic ecology, and estimations of biomass and energy flow (Ryan, 1977; Block, 1977; Procter, 1977). Though many polar invertebrate species appear to be endemic, very little is known of their biology: the liveliest, most prominent and most numerous elements are the arthropods, and these have provided most of what is known on adaptations and ways of life in terrestrial and freshwater environments.

6.3.1 *Insects*

Many non-polar insects that live in northern continental areas are subject to extremes of cold in winter and heat in summer; alpine insects may meet similar conditions every day. In some species there is evidence of supercooling, i.e. failure of the body fluids to freeze at temperatures appropriate to their osmolarity. In others, dehydration and the accumulation of glycerol or other antifreeze agents in the body fluids appear to be implicated in lowering the freezing point (Danks, 1978, 1981). Species vary in their ability to tolerate freezing. Some die at the first hint of frost; others, for example the damp-living larvae of chironomid midges, can be subjected repeatedly to temperatures of −20°C and below, with up to 90% of their body water frozen, and recover completely on thawing (Scholander *et al.*, 1953).

Adaptations of arctic insects are summarized by Kevan and Danks (1986a and b) and Chernov (1975, 1978, 1985). Overall these species are darker, smaller and hairier than temperate counterparts. Darkness and hairiness, especially noticeable in butterflies of the genera *Boloria* and *Colias*, are taken to be adaptations for absorbing and retaining heat during basking. Smallness facilitates rapid warming to working temperatures, though it may also be a symptom of food sparsity; compactness, including relative shortness of limbs and wings, helps to conserve heat and moisture. Wings are short or absent in several arctic species of stone flies, crane flies, midges and moths; females are especially likely to have reduced wings and to avoid flying. Eyes and antennae are relatively small in some species, suggesting that sensory inputs are limited.

All insects are most active in summer; though seldom apparent on cold, sunless days, those of the high tundra must be able to hunt, feed, evade predators and reproduce at temperatures which would immobilize species from lower latitudes. Only bumblebees in flight are known to maintain higher-than-ambient body temperatures; heat is generated by the flight muscles and kept from escaping by the furry surface of thorax and abdomen.

Basking is highly characteristic of arctic insects. Many diptera bask in sunlight inside flowers, warming themselves and at the same time facilitating pollination; hordes settle on conspicuous dark or bright surfaces when air temperatures are low, and beetles collect on sunny patches of ground that are sheltered from wind, indicating a constant need for warming. Flying insects keep close to the ground and out of the wind, haunting sheltered valleys and lees of hills, seeking direct sunlight as much as possible. Swarms and mating flights seldom rise more than two or three metres off the ground; aerial mating may be replaced by mating on surface vegetation.

Life cycles. Life cycles are adapted to low temperatures. Where similar temperate species take a year or less to complete their life cycles, polar species may take several years. Individuals feed and grow slowly, spending longer at each growth stage and ultimately producing fewer eggs. All stages from eggs to adults must be capable of withstanding low environmental temperatures; eggs especially have to be cold-resistant throughout the winter.

Overwintering strategies of many species are unknown. Of those that are known, aquatic insects that winter as immature stages in large ponds probably have the warmest environments. Eggs, larvae and pupae remain unfrozen under the ice at temperatures between 0° and 4°C. Those in smaller ponds must experience much lower temperatures when the water about them freezes. Many aquatic flying insects stage mass emergences in early spring, suggesting that winter dormancy is a necessary part of their cycle. Sex is often unnecessary; a high proportion of polar insects including caddis flies, stone flies, midges, simuliid black-flies and plant bugs are parthenogenetic. A few are polyploid (cf. plants, above).

Insects that winter on land, under snow, among rocks and vegetation or in soil are similarly exposed to extreme cold. Blowflies, psillid bugs and bumblebees are among the few that winter as adults, seeking sheltered corners under rocks and vegetation, close to the soil and under an insulating layer of snow. Blowflies winter also as maggots or pupae in carcasses; butterflies and moths probably winter only as larvae or pupae. Dehydration and the accumulation of glycerol in tissues facilitate freeze resistance at all stages of life cycles from eggs to imagos (Danks, 1978, 1981).

Antarctic insects and mites have been widely collected from stations that include some of the world's southernmost outcrops. Collembola live among mosses, lichens and turf, feeding on fungal hyphae, dead plant material, live algal cells and microorganisms. A few are present on almost all vegetation, wherever it is found. High densities of up to 100 000 m^{-2} have been recorded on moss turf in the South Orkneys, and on warm summer days ponds in South Georgia are occasionally covered with a grey film of collembola, apparently newly emerged from the soil. Of 47 species that occur in the Antarctic region, Wallwork (1973) considers about 70% endemic; 17 species occur in continental and maritime Antarctica. Rapoport (1971) characterizes antarctic species as heavily pigmented; otherwise they appear unremarkable, with few special adaptations for polar life.

Their basic ecology, whether studied on the continent (Prior, 1962; Wise and Shoup, 1967; Janetschek, 1970) or on the maritime and periantarctic islands (Strong, 1967; Etchegary and others, 1977; Tilbrook, 1970, 1977; Block, 1982) is similar throughout their range, indicating that collembola, like lichens, are well adapted for harsh environments but not especially equipped for polar life. The upper lethal temperature for one species, *Gomphiocephalus hodgsoni*, was as high as 29.5° in moist conditions; the lower lethal temperature ranged between −11°C and −23°C, unaffected by acclimation. In *Cryptopygus antarcticus*, the subject of intensive studies by Tilbrook (1977) and Block (1982), specimens from South Georgia were on average smaller and lighter in weight, and matured at smaller size, than those from Signy Island, but respiration–weight and metabolism–temperature curves of the two stocks were similar. Cold-adapted collembola may metabolize slightly faster than temperate species; in laboratory stocks of *C. antarcticus* moulting occurs down to lower temperatures than in temperate stocks (Burn, 1981).

Very little is known of the ecology or adaptations of periantarctic beetles, grass aphids or thrips. Two species of diptera, both chironomids, occur in the maritime Antarctic. *Belgica antarctica* (Figure 3.12), a wingless midge, is confined to the shoreline of northwestern Antarctic Peninsula and the South Shetland Islands, where its larvae feed in moist soil and tide pools. Adults are not frost-tolerant; only eggs (and possibly larvae) survive the winter. *Parochlus steinenii*, a winged midge, occurs sparsely about the South Shetland Islands and more commonly on South Georgia.

6.3.2 *Freshwater and marine invertebrates*

Heywood (1984) typifies the fauna of antarctic lakes and ponds as non-specific opportunists, feeding on epiphytes, detritus and bacteria, that encyst or produce resting eggs when conditions become severe. Rotifers have for long been know to survive repeated freeze-thaw cycles. Among antarctic crustaceans the anostracan *Branchinecta gaini* overwinters as eggs, which hatch in spring, possibly in response to osmotic changes. Those encased in ice hatch with the thaw; those already in water hatch as the medium surrounding them is diluted by melt-water.

Adults grow to varying sizes, depending on food available, and lay eggs in late summer. The adult population dies when an ice cover forms and oxygen levels within the water fall, after a lifespan of about three months in ponds and up to six months in lakes. Copepods *Parabroteas sarsi* and *Pseudoboeckella poppei* overwinter only as eggs in water bodies that freeze completely or become anoxic; in large lakes they are active year-round, breeding in late winter to produce nauplii that feed on the early plankton bloom. Resting (i.e. overwintering) eggs appear from midsummer onward.

Marine invertebrates live in more stable environments than those of land or fresh water and generally face less extreme conditions; they are unlikely, for example, to dry out, freeze, or suffer wide changes of temperature or salinity. Their body fluids are isotonic with sea water, and they have no special problems of freezing comparable with those of fish (see below). Adaptations of intertidal animals in cold climates have been studied by Kanwisher (1959). Generally rare in polar regions, where sea ice inhibits their settlement, they are often plentiful on subpolar shores where they may be exposed twice daily to six-hourly spells of near-freezing sea temperatures, alternating with similar spells of very much lower air temperatures. Those examined by Kanwisher simply froze when exposed to extremely cold air. At −30°C up to 90% of their body water became ice, which formed mainly in intercellular spaces, resulting in high but apparently non-lethal concentrations of solutes within the cells.

6.4 Survival in fish, birds and mammals

6.4.1 *Fish*

Of the 20 000 or so living species of fish, only about 120 are found south of the Antarctic Convergence, and most of those are endemic (Everson, 1984). The Arctic too has a small though less exclusive fish fauna, but there are no known bipolar species. Most polar fish live in waters at temperature that are very stable and close to 0°C. Low temperatures place them at no obvious disadvantage—they have evolved physiological systems that allow them to move just as fast as temperate or tropical fish, and they are no easier to catch.

Possibly resulting from low-temperature adaptations, they tend to grow seasonally and overall more slowly, take longer to mature, and produce fewer, larger eggs (Everson, 1987). They are mostly stenothermic, i.e. unable to survive temperature ranges of more than a few degrees, or even to acclimatize effectively. The antarctic shallow-water marine fish *Rigophila dearborni*, accustomed to living at −1.5°C, dies quickly at temperatures only 0.5°C lower or 4–5°C higher (DeVries, 1978).

The body fluids of marine fish are generally more dilute than those of sea water, and would be expected to freeze at temperatures of −0.6° to −0.8°C (Schmidt-Nielsen, 1975), a degree or more higher than the freezing point of polar sea water. Body fluids of some polar fish are slightly more concentrated than those of temperate species, but ionic differences in osmolarity are minute and have little effect on freezing point. Non-pelagic species, which do not normally come into direct contact with ice, appear to avoid freezing simply by supercooling; their fluids remain ice-free even 1–2°C below freezing point. If touched by ice experimentally their gills and other vulnerable tissues freeze, and they die immediately (Scholander *et al.*, 1957).

Pelagic species that live among ice contain antifreeze agents, often unusually high concentrations of plasma proteins (Hargens, 1972). One agent occurring in the antarctic species *Trematomus borchgrevinki* (Figure 6.3) has been identified as a glycoprotein of molecular weight up to 21 500, which appears to bind itself to ice crystals and inhibit their

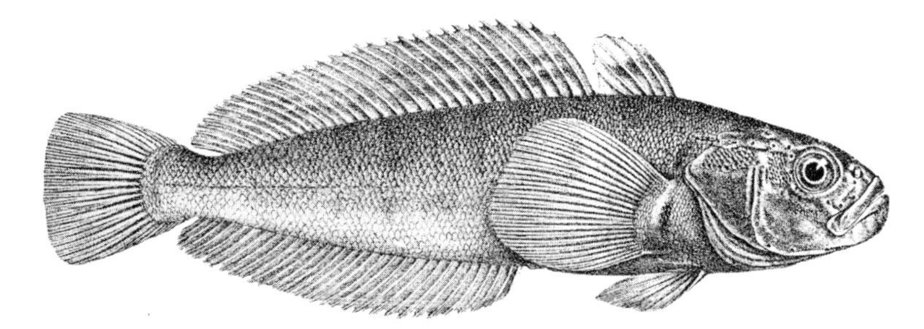

Figure 6.3 *Trematomus borchgrevinki*, an antarctic inshore pelagic fish; length 30 cm. Its blood contains antifreeze, allowing it to feed in water where ice crystals are present (Boulenger, 1902).

growth (Lin *et al.*, 1972). Weight for weight it is far more effective than salts in depressing freezing point, and physiologically far less intrusive on cellular activities. Presence of these substances in the blood presumably replaces any need for supercooling, and allows pelagic fish to hunt in the highly productive region immediately under the ice floes.

Channichthyid fish of antarctic waters, for example the icefish *Champsocephalus aceratus*, are unusual in possessing only a few erythrocytes (red blood corpuscles) as larvae and none as adults; related species have much-reduced erythrocyte counts throughout life. Their oxygen requirements are normal and blood systems are similar to those of other fish, except that the heart is larger and gill blood vessels are comparatively wide, allowing a faster circulation of blood. These fish appear to rely on a rapid circulation of plasma to meet their oxygen needs; what advantage (if any) this strategem confers is not clear, but it is one that would not be feasible in warmer oceans where the oxygen content of the water is lower.

6.4.2 *Homeotherms*

Warm-blooded animals the world over maintain a high and constant core temperature. Their main source of heat is their own metabolic activities, and nearly all maintain core temperatures of 38.5°C (most mammals) to 41°C (most birds). Assuming a mean temperature of 14°C for the earth and lower atmosphere as a whole, all but a very few homeotherms in tropical regions lose heat constantly to the environment. Losses are reduced by insulation and behaviour, but must ultimately be made up from metabolic energy sources, mostly by cellular oxidation of fuels derived from food. The benefits of homeothermy are apparent from the wide variety of form and lifestyles adopted by birds and mammals, but the costs are high. A mammal or bird of given weight in a temperate climate requires between three and seven times as much food to keep going as a reptile of similar weight.

Both the environmental temperatures and the body activities that provide the main source of internal heating fluctuate from minute to minute, yet the point of balance represented by core temperature remains remarkably constant. Very precise feedback control mechanisms ensure that, though skin temperatures may vary considerably, core temperatures remain steady within 1°C or less; for useful discussions of how this is achieved, even in polar environments, see Kleiber (1961) and Schmidt-Nielsen (1975, 1976).

In pioneering studies of Alaskan birds and mammals from a wide range of environments, Irving and Krog (1954) concluded that polar homeotherms maintain core temperature similar to those of their temperate and tropical counterparts. Though many more polar homeotherms have been examined since then, evidence to the contrary has been slight. Except in rare cases of torpor or hibernation (neither is generally used in polar environments), homeotherms in cold climates show no special adaptations involving temperature control.

In fact polar homeotherms, whether northern or southern, show very few recognizable adaptations for polar living. Many appear to be recent colonists from temperate regions, representing species that have extended their breeding ranges northward or southward in post-glacial times. Carrying their own temperature environment within them, their adaptations, such as they are, relate mainly to seasonal cycles of food. Most avoid extremes of polar cold and dearth of food by wintering elsewhere, and using the polar environment only during its most tolerable phase. Species that overwinter make use of their avian or mammalian pre-adaptations, rather than special modifications, for polar life.

The few year-round species of high-arctic birds and mammals make good use of feathers, fur and subcutaneous insulation. Small mammal species (voles, lemmings) dig or burrow in snow, isolating themselves from environmental extremes and keeping in touch with food at the ground–snow interface. Larger grazing species (caribou, musk oxen) are especially well furred. Taking as a practical measure of insulation the 'clo' unit (equivalent to normal clothing worn by a human at rest in an ambient temperature of 20°C), wolves and polar bears carry insulation of 6–7 clo, and arctic foxes in winter carry 8 clo, rather more than is deemed efficient for human polar clothing. Large mammals also select their environments carefully, and bunch together to reduce surface heat losses. The whiteness of polar bears and arctic foxes provides camouflage, and may also allow solar radiation to penetrate and heat the skin surface directly.

Aquatic birds and mammals, equipped with subcutaneous blubber, may also have a covering of fur or feathers. The waterproof feathers of many marine birds, underlain by a dense blanket formed from the aftershafts of individual feathers, enhance insulation and serve also to keep out wind and snow. Polar penguins are slightly better insulated against their extreme environmental temperatures than are species of temperate seas (Stonehouse, 1967); temperate species need good insulation, and that of polar species need be only slightly better to be effective in lower temperatures. Emperor penguins (Figure 6.4), breeding furthest south and across midwinter, not surprisingly possess the best insulation of all.

Whales and seals, the largest polar aquatic mammals, rely almost entirely on blubber for insulation. Whales that migrate annually to polar seas enter

Figure 6.4 Emperor penguins: part of a breeding colony of over 11 000 pairs in 76°S. This species incubates through the Antarctic winter. Photo: B. Stonehouse.

the cold regions in early spring with very thin blubber, and fatten markedly during the polar summer. Fur, which relies on trapped air for insulating efficiency, cannot be effective in animals that spend all their lives in water, and rapidly loses effectiveness by compression in those that dive deep. The thin pelts of hair (phocid) seals and denser pelts of fur seals probably afford better protection against wind-chill than against the cold of the sea (Figure 6.5).

Special adaptation is required in a few species to fit breeding cycles into the polar year. Most polar birds breed seasonally, mating in early spring and timing their laying and incubation so that chicks are fed and released

Figure 6.5 A breeding group of Weddell seals basking in Antarctic summer sunshine. Thinly furred, they rely on 3–4 cm of blubber for insulation. In the cold windy weather of winter they spend most of their time in the sea, where conditions (at $-1.8°C$) are apparently preferable. Photo: John Darby.

during the summer flush of food. Among the larger ones, Adélie penguins *Pygoscelis adeliae*, living close to Antarctica at the southern edge of their range, start to breed in October or November (early spring). By doing so they just manage to have their chicks ready for sea by late February, and complete their post-nuptial moult before the March freeze-up. Still-larger emperor penguins *Aptenodytes forsteri*, with longer incubation and chick-rearing periods, set back their courtship and laying to June (mid-winter), incubate through the coldest months of July and August, and rear their chicks through early spring. The chicks are released precociously at only about half adult weight, giving the adults time to complete their own moult and be ready for breeding the following season (Stonehouse, 1985). For seasonal breeding and feeding in other southern birds see Siegfried (1985); for accounts of northern birds and mammals see Sage (1986).

6.5 Summary and conclusions

Polar species are recruited from neighbouring regions; the wider variety living on land within the northern polar boundary reflects wider subpolar and temperate recruiting grounds in the north than in the south. Relatively few species using polar regions are endemic. Several categories of plants and animals are preadapted by life in temperate region for an easy transition to polar regions; they need to change little once they are established, so speciation pressures are light. Polar cold may be less contricting than other factors, for example aridity or intense seasonality, that characterize polar regions. On land lichen, mosses, many flowering plants, mites, insects and other invertebrates, and many species of birds and mammals, come into this category. Aquatic mammals and birds also are pre-adapted by life in cold temperate waters, and seem well able to cope with the marginally colder conditions in polar seas. Acclimation and adaptations for polar living are discussed, and the life-cycles and physiological traits of some polar organisms are examined.

CHAPTER SEVEN

MAN AND THE POLAR REGIONS

7.1 Introduction

Evolved in the tropics, most naked of all the apes, man is physiologically among the least likely of all warm-blooded creatures to flourish in polar regions. Yet from the tropics he spread in stages northward and south into cooler regions, reaching the Arctic—a milder Arctic than the present one—about 30 000 years ago. Later generations spread south as far as the tip of South America; later still they reached similar latitudes in southern New Zealand. Vagrants from these populations may have explored the Southern Ocean and landed on some of the cool temperate islands, but there is no convincing evidence for their reaching Antarctica. Man unclothed would not have survived long in the far north or south. Only by wearing the furs of other animals, by use of fire and by liberal application of his wits, has northern man has survived as an indigenous polar and subpolar animal for more than 20 000 years.

Human problems in extreme cold are both physiological and environmental. Polar folk, like other homeotherms, show little evidence of special physiological adaptations for life in cold climates, and the meagre environmental resources have kept their populations small, widely dispersed and nomadic. In recent years new problems have arisen from the mixing of indigenous polar cultures with those from outside, introducing new ecological demands and imposing new stresses on people and environments ill-equipped to cope with them.

7.2 Man and cold

7.2.1 *Homeothermy*

Like other homeotherms, man maintains a high and constant body temperature against the environmental gradient. He does so by metabolizing fuels in muscular and other body cells. At rest in a mild environment his basal or standard metabolic rate is about 2000 kcal or 8400 kJ per 24 h, equivalent in work units to just less than 100 W, the output of a reading-lamp bulb.

Unclothed and at rest, cold-adapted humans can maintain a core temperature (i.e. central body temperature, measured by mouth or rectum) of 37°C in still air down to 27–29°C, or in water down to 35°C; in either case the skin and extremities are likely to be several degrees cooler (Edholm, 1978). As core temperature starts to fall, self-regulating mechanisms start to restore equilibrium. Shivering, a form of involuntary muscular action, raises the metabolic rate and elevates body temperature; in extreme cases it can quickly raise metabolic rate by a factor of four to five. To provide this energy, stores of blood sugar and fats are metabolized.

Metabolic rate rises with all other forms of activity, by factors of two or three in walking and up to eight or ten in more strenuous exercise (Durnin and Passmore, 1967). Fuel consumption rises proportionately and excess heat is likely to be generated. For body temperature to remain stable, the excess heat must be shed in exhalation and by radiation, conduction, evaporation and convection from the body surface. Not surprisingly, food requirements increase substantially with activity.

Infants enhance heat production by metabolizing brown fat, a form of storage tissue, mostly subcutaneous, rich in mitochondria and biologically active. Though used by the young of many homeotherms, this method of warming is not open to adults, even those of populations subject to extreme cold, for the brown fat disappears before adolescence. Other forms of 'non-shivering thermogenesis' may be used by human populations that encounter cold. Alacalufe Indians of Tierra del Fuego, for example, maintain a slightly enhanced metabolic rate without shivering while sleeping in cold, damp conditions. After days in searing heat, Australian aboriginals withstand cold nights in the desert without shivering; possible mechanisms are enhanced metabolism, a lower threshold for the start of shivering, and a slight overnight fall in core temperature.

7.2.2 *Man in polar environments*

No special adaptations for dealing with cold, beyond the level of acclimation, have been found in indigenous Arctic folk. The squat, short-limbed build once characteristic of the Eskimo (now generally called Inuit) is rapidly disappearing in modern generations (Godin and Shephard, 1973); it was probably as much a product of malnutrition as of genetic endowment. This build would ensure the smallest possible surface area through which to lose heat, though the effect in well-insulated subjects can only have been trivial. Inuit may have slightly higher than average basal metabolic rates (though this too is doubtful; *ibid.*) and lower threshold for onset of shivering.

Those who pursue the traditional hunting way of life tend to be lean and fit (Shephard and Rode, 1973). Like fishermen and others who work in cold conditions, they develop tolerance to cold in hands and faces, possibly a function of enhanced blood circulation in exposed areas of the body. However, the success of the Inuit is based far more on dressing, housing and feeding themselves as well as possible, and avoiding risks that would expose them unduly to cold. Similar strategies are adopted by the non-indigenous people who leave temperate environments to live temporarily in polar regions; they make small physiological adjustments, but rely for safety on good clothing, housing and food, and constant alertness to risk.

Metabolic activity arising from muscular exercise, coupled with insulation in the form of layered windproof clothing, is generally enough to maintain human body temperature against polar extremes. Insulation up to 3 clo units (p. 177) is effective for active people, and up to 4 units for sedentary work (Brotherhood, 1973); additional thickness of clothing tends to be cumbersome and counterproductive, requiring more energy to carry it about. Heat is readily generated by activity; even increased muscular tone generates enough to restore comfort in chilled but well-insulated subjects. Running, climbing and other strenuous action soon produce a need to shed heat, indicated by raised skin temperature and the onset of sweating. This can often be balanced by temporary exposure of skin, e.g. lowering an anorak hood or removing one or both gloves.

To avoid condensation and accumulation of sweat, with consequent risk of chilling and internal frosting, polar clothing must be well enough ventilated to allow vapour to escape during activity. Inuit skin clothing is loose-fitting with built-in ventilation; for modern polar wear, much thought has gone into the design of garments that are lightweight, comfortable, windproof, and efficient, whatever the wearer is doing.

Wind is notorious in making cold seem colder, chiefly in its effects on exposed skin. 'Wind-chill factors', calculated originally in experiments involving heat losses from dry uninsulated surfaces (Siple and Passel, 1945), are often represented either as reduced temperatures or as additional heat lost per unit of time. Both indices can be misleading when applied to real situations involving man or other homeotherms. Well-insulated subjects in windproofs (similarly, birds or mammals with wind-resistant plumage or pelage) may be virtually unaffected even by very strong winds. By contrast, poorly-insulated subjects, especially if wet, have their rate of cooling dramatically increased by even light winds that are strong enough to disturb the boundary layer around them. Thus an inadequately-clothed walker or climber may be at risk if exposed to winds to 2–3 m per second, and only marginally more at risk in winds three or four times as strong. In either case safety lies in finding either a windproof covering or a sheltered corner as soon as possible. The surface of a warm, damp body takes up the wet-bulb temperature of the air around it. Wind may secure this equilibrium faster, but no wind, however strong, can reduce the temperature of a body below its own wet-bulb temperature.

In traditional skin clothing or modern padded polar suits, man can work outside without difficulty at environmental temperatures down to −40°C, and with extra precautions (face masks, electrically-warmed suits) at temperatures considerably lower. On the Antarctic polar plateau in high summer at −36°, Dalrymple (1966) reported low wind speeds, little cloud and strong sunlight, combining to create 'a rather pleasant environment'. While Siple (1959) and his colleagues (including a dog) strolled comfortably about the South Pole in calm weather with temperatures of −70°C, prolonged work outside became difficult, even dangerous, below −55°C.

More extreme cold is tiresome as well as dangerous. Ignatov (1965) reported that in winter at Vostok, almost 3500 m above sea level on the high Antarctic plateau, kerosene (paraffin) becomes a jelly at −60°C and solidifies at −85°C. At −75°C diesel fuel has to be chopped with a hatchet. About −80°C, antifreeze turns to pink ice and gasoline (petrol) fails to vaporize or ignite. Scientists who have worked on the high plateau forgivably take a cavalier view of coastal and maritime Antarctica, writing it off as 'that banana belt up north'.

Inhaling very cold air may cause local freezing in air passages and lungs, or spasms similar to asthma that make breathing difficult. At all subzero temperatures there is a some danger of frostbite (local freezing of tissues in the face or extremities). This may lead to blistering, and, if untreated, to necrosis and gangrene. However, life in polar regions is generally healthy, strenuous and safe. Paradoxically, danger and discomfort are often severest at temperatures close to freezing point, when outer clothing, wetted by rain or melting snow, loses some of its insulating properties. Deaths due to exposure are more likely during winter in temperate mountains and uplands than on properly-managed polar expeditions.

Those who live in cold climates require high-calorie diets, but only if they are living actively in the open air; sedentary indoor workers need no more than their counterparts in temperate latitudes. Activities associated with exploration—sledging, ski-marching, hunting, tending dog teams, even walking on snow and carrying heavy clothing—all involve hard work, for which two to three times the normal caloric intake is required. The additional calories are spent on muscular activity, and the heat generated is more than enough to maintain the subject's temperature against the steep environmental gradient.

7.3 Arctic man

People first reached Siberian Arctic coastlands from central Asia, successive waves moving northward to populate what may then have been a relatively mild maritime area. Man was among the last of a wide range of Eurasian mammals to trek overland through the forests and tundra of Beringia, now the Bering Strait region, during the late Pliocene and Pleistocene. Some 10 000 years ago, detachments began to move eastward from Asia to North America. From there, man spread eastward and south, ultimately to occupy the whole of the new continent to the southern tip of South America (Hopkins, 1967). As the last (Wisconsin) ice sheets advanced in the north, human populations shifted southward before them. As the ice retreated the northlands again became tenable, and some 6000 years ago coastal populations from northeastern Siberia spearheaded new waves of immigration eastward.

The first American polar folk were well clothed in sewn animal skins and equipped for maritime life with kayaks (skin boats), fish-hooks, harpoons of bone and ivory, bows and arrows. A shamanist philosophy helped to even the odds against them, and they spread through the Aleutian Islands and along the Arctic coast of North America (Figure 7.1), forming the distinctive northern folk formerly known as Eskimo, now called Inuit in recognition of their pan-Arctic racial identity. Successive waves spread through the Canadian archipelago to northwest Greenland, moved down the Mackenzie River, and crossed eastward to Labrador and southern Greenland.

7.3.1 *Inuit, Saami and others*

Until very recently the Inuit, the true Arctic folk, lived as their ancestors had lived during those early days of coastal occupation, in small, nomadic groups with rich culture but few material possessions (Birket-Smith, 1959; Herbert, 1976). Physically small, compact people (but see above), they kept separate from their nearest neighbours, the North American forest Indians,

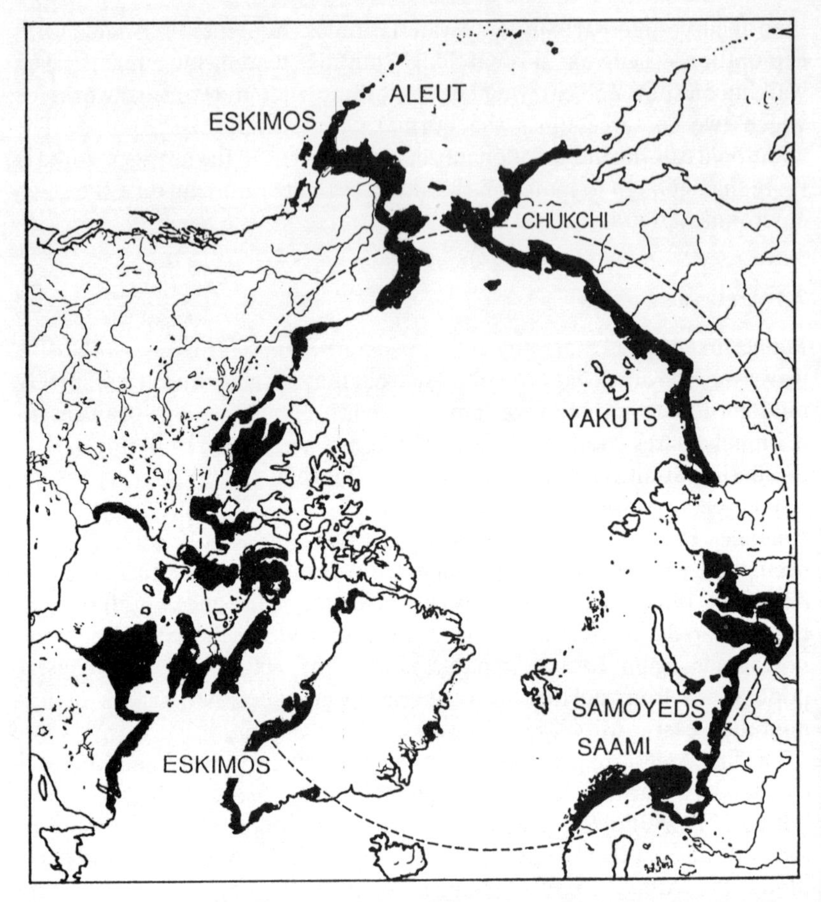

Figure 7.1 Arctic coastal regions and their indigenous peoples. Eskimos have tended to remain coastal and estuarine. Eurasian populations generally live inland, close to rivers and lakes.

maintaining a common northern culture varied by local customs and practices. Their way of life involved annual circuits from one seasonal hunting ground to another, using the few materials available to them—stone, clay, bone, skins, sinews, and a little driftwood timber—with dog sledges for transport.

Traditionally in summer they were based ashore, living in skin tents and sod huts, hunting and fishing from skin canoes (kayaks and family-sized umiaks) among the ice floes, trapping fish at weirs and stone barriers in the rivers, and searching the tundra for caribou, musk oxen, smaller game, birds, eggs and berries. In winter some continued ashore, reinforcing their tents and huts for extra warmth, while others moved out on to the inshore sea ice, living in snow houses and feeding mostly on fish and seals.

Arctic folk of Europe and Asia—the Saami (formerly called Lapps) of Scandinavia and the northwestern USSR, and the Nenet, Yakut and Chukchi of Siberia—share common ancestry with the Inuit and are physically similar, but have developed more land-based ways of life. Only those in the far east, bordering the Pacific Ocean, were truly maritime; the rest centred their economies on river fishing, hunting, trapping and reindeer herding. These patterns of life were still widespread at the start of the 20th century; a few pursue it today, but under the influences of civilization, most have settled and traded at least part of their culture for southern ways.

7.3.2 *Exploitation*

Protected by their icy environment, northern folk were for long virtually isolated from southern influences. The skills and cultures they developed sustained them in a harsh living, and gave them a lasting, non-destructive role in Arctic ecology. Their small nomadic populations, growing and subdividing during the good years, were periodically culled by cold and starvation. Indigenous human populations only marginally affected the huge northern stocks of whales, fur seals, seals, walruses and fish. With little access to trade outlets, early Inuit had neither the power nor the will to over-exploit or destroy their few resources. Mediaeval explorers from the south made little impact on their society, but incursions of southern whalers, sealers, missionaries, fur traders and administrators on an unprecedented scale during the 18th and 19th centuries exposed them more fully to southern civilization, often with devastating results (Hall, 1987).

Hunters from the south who took thousands of whales and seals did not generally compete with the indigenous folk or reduce the food available to them. Their most damaging impacts were the introduction of diseases and alcohol addiction, which incapacitated and occasionally destroyed whole communities (*ibid.*). Fur trading had a more insidious ecological impact, for what the indigenous people had taken on a small scale for subsistence was now demanded on a large scale for sale in the south. The acquisition of steel knives, guns and patent traps made hunting and everyday life easier for the Inuit, but the price paid in furs for these trade goods represented over-exploitation of a meagre community resource. From intensive trading arose changes in the life style, values and population structure of local communities, massive over-use of resources, and the need for new economic inputs to maintain northern communities at even a minimal standard of living.

By the late 19th century many northern populations were decimated; the survivors lived half-starved in squalor and disease, unable to continue their traditional lifestyles but without viable alternatives. Their plight was gradually recognized during the early 20th century; from the mid-century onward it has been the role of administrators, teachers, health workers, economists and lawyers to unravel some of their problems and provide alternative means of living. Many of those responsible saw their task as one of assimilation—to bring northern folk into southern-based cultures and modern ways. More recently, northern folk have seized a greater share in responsibility for their own affairs; they seek to take advantage of southern benefits, while at the same time maintaining their northern identity.

The most humane and ecologically sound management policies, made possible by constant inputs of money from the south, combine elements of Inuit culture with the elements of southern life that are most useful to them. Thus a modern Inuit village family may live in a small settlement of insulated timber housing, spend part of their time working for wages in government or private company employ, hunt caribou or seals for meat, and tend traplines for extra money to buy consumer goods from the store or mail-order catalogue. They are today more likely to wear store clothing than skins, and to travel by skidoo (motor sledge) than dog team. Village schooling and primary health care will be available locally, with more advanced education and medical services in the main centres.

The basic ecological problem of limited resources remains. The biological resources that just supported small nomadic populations at a neolithic level of culture cannot be expected to support larger populations at southern standards of living. Agricultural development is restricted to the subpolar fringe and generally unpromising. Though crops can be raised for local consumption in favoured areas of Alaska and Siberia, efficient horticulture is possible only under glass and with considerable inputs of energy; it is generally cheaper and more reliable to bring in produce by road or air from the south. Only a few marine biological resources are capable of massive exploitation, but these generally require more capital to develop than can be generated locally. The fisheries developed off Greenland and Iceland, started by colonial investment and now protected from foreign exploitation, are helping to raise standards of living ashore.

Non-biological sources of income, including revenues from defence installations, scientific surveys and mineral exploitation, are also helping to keep northern communities viable. However, the high costs of bringing building materials, fuels and plant to the Arctic, and the difficulties and expense of exporting ores and other minerals to where they can be used, discourage development of all but the most urgently needed mineral resources; at present oil is the major export from Arctic North America. Tourism is a slowly-growing source of revenue, helped by the development of air transport networks and the dedication of parks and wilderness areas.

Thus northern communities that live above subsistence level generally have to be subsidized, and can expect to survive only while subsidies last; their continuance depends on political decisions, which vary from one part of the north to another (Armstrong and others, 1978). Under Soviet rule, Asian indigenous communities have largely been assimilated into southern ways. In North America and Europe, divided politically under United States, Canadian and Greenlandic rule, the Inuit are both reaping the benefits and suffering the inconveniences of a rapid culture-shift from traditional to modern ways. They retain a common language and are developing a political awareness that may ultimately unite them into a distinct western Arctic nation.

7.4 Man in the Antarctic

Antarctica has no indigenous population. The Southern Ocean was virtually unknown to man before the 16th century, apart from possible visits to the ice edge by wandering Polynesian mariners. The southern islands became known mostly during the 17th to 19th centuries; Antarctica itself remained unvisited until the late 18th century, and unrecognized as a continent before the start of the 20th century. Biology played an important role in southern exploration, for many of the early explorers were sealers and whalers. The current population is mostly made up of scientists and support staff working on contract for one or two years. Though several nations claim parts of Antarctica, virtually no-one regards the continent as home.

7.4.1 *Discovery and exploitation*

Though geographers of classical times had predicted the existence of a southern polar continent, exploratory voyages of the 16th and 17th centuries ventured no further than the temperate zone of the southern hemisphere (Mill, 1905). The ice-covered ocean surrounding Antarctica was first penetrated by Captain James Cook, RN, in his exploratory voyages of 1773 and 1774 (Cook, 1961). Of more immediate significance to contemporaries were his observations of fur seals on southern islands, especially South Georgia. From the late 18th century onward these brought sealers from

North America and Europe into the Southern Ocean; during the subsequent decades of competitive exploitation it was sealers who discovered most of the cold temperate islands, the ice-covered islands within the pack ice belt, the tip of Antarctic Peninsula, and the great ice-filled bight of the Weddell Sea.

Major Russian, French, British and US expeditions of the 1840s made further discoveries in the far south, effectively defining the shores of the predicted continent almost to their present positions (*Reader's Digest*, 1985). Reports of an abundance of whales in the Southern Ocean brought European whalers south during the late 19th century, when scientific exploration of the continent also began (Hayes, 1932). Geological exploration both coastal and inland yielded reports of coal measures and mineral ores, though in small amounts and generally difficult of access.

Unsuccessful at first, southern whaling began to flourish from the first decade of the 20th century, based initially at shore stations on South Georgia, later on pelagic fleets of factory ships and catchers. These pursued the large whales, mostly rorquals and sperm whales, penetrating southward into the pack ice and almost to the shores of the continent. In a strongly competitive industry, high levels of skill and technology brought success for over half a century. The industry continued on a large but declining scale until the 1960s, by which time most of the big rorquals had been hunted beyond economic levels. Today there is very little hunting in Antarctic waters, and stocks of the larger whales remain sadly depleted.

From the 1920s onward, efforts were made to rationalize and control southern whaling, notably through the International Whaling Commission; protective measures included prohibitions on the killing of the most heavily depleted species, protecting immature whales and females with calves, and the introduction of seasonal quotas. These measures were more for the maintenance of the industry than to save whales. However, not enough was known of the biology of stocks to make control measures credible or effective, and whalers in international competition for declining stocks had no alternative but to press for the largest allowable quota, and hunt at maximum efficiency to attain the largest share. For a review of the decline of southern whaling, see Bonner (1980).

More recently, pelagic fleets of several nations have fished the Southern Ocean effectively for pelagic and demersal fish, squid and krill; for overviews of their catches and potential see Grantham (1977), Eddie (1977), Everson (1977) and Knox (1983).

Agriculture in the south has been restricted to stock-raising, mostly for the local consumption of whaling communities, on some of the peripheral and cool temperate islands. On both South Georgia and Iles Kerguelen, free-ranging reindeer have proved especially successful. Populations of smaller mammals, including rabbits and rodents, have also been established in these latitudes, generally to the detriment of the habitat (Bonner, 1984); none has yet survived on mainland Antarctica or on islands within the pack ice belt.

Mineral and other non-biological resources are present but have not so far been exploited; for surveys of mineral potentials, see Holdgate and Tinker (1979) and Gjelsvik (1983).

7.5 Polar conservation

Taking 'conservation' to mean the wise use of resources, this section explores current strategies for the conservation of polar resources at either end of the world.

Though barely habitable by man, the polar regions have for generations yielded goods that are valued by man. The Arctic's small indigenous populations live at subsistence level, or are subsidized, like its non-indigenes, toward higher standards of living; Antarctica has no indigenous humans, only floating populations kept there at considerable cost for scientific or political ends. Except for marine products, the few goods that polar regions yield are seldom valued highly enough to offset the high costs and risks of exploiting them.

Much of the northern produce has come from the subpolar maritime fringe, notably whalebone and whale oil from the open sea, seal oil, hides and furs from islands and sea ice, fish, and, most recently, shrimps from the continental shelf and slopes. Boreal subpolar forests have yielded furs, and northern lands and continental shelves are currently being explored or exploited for minerals, including coal, metallic ores, oil and gas. Products and yields vary according to fashion and demand; whaling, sealing, fur trading and mineral ore exploitation have declined in recent years, but commercial fishing and fossil fuel exploitation are well established and likely to continue.

Southern polar products are entirely maritime; the Southern Ocean, especially on its subpolar fringe, has yielded whale and seal products, and is currently being exploited for fish, squid and krill. No minerals have yet been exploited in Antarctica or the subantarctic islands, though coal and metallic ores are known to exist in exploitable amounts, and there are strong expectations—as yet no more—that oil and gas may occur in exploitable deposits offshore.

The term 'resource' covers all assets available to man, and some further resources of polar regions are coming to be valued increasingly in a crowded world. Both regions include splendid scenery and wildlife, for which a possible wise use is well-managed tourism. Both include wilderness, an altogether more delicate asset that is to some degree marred by almost any human incursion. There is no wise use for wilderness, except to preserve it by excluding man altogether. A working compromise is to allow strictly regulated rights of passage, that permit monitoring but exclude all possibility of exploitation in any other way.

7.5.1 *Overexploitation*

From the 16th to the 19th century, in some places even into the present century, exploitation of polar resources has been largely unmanaged, and limited only by market satiation. Whaling on the high seas, like right of passage, was traditionally free for all; so were commercial sealing and fishing. As new lands were discovered, rights to exploit inshore fish, seals and other assets were sometimes claimed along with the territory, but neither policing nor management were usually practicable. Uncontrolled competition, hungry markets and high profits during the 18th and 19th centuries virtually destroyed northern polar and subpolar stocks of whales, fur seals, elephant seals, sea lions and sea cows. During the 19th century Southern Ocean stocks of fur seals and elephant seals were similarly reduced almost to extinction. Simultaneously many southern islands were devastated by the introduction of pigs, goats, sheep, cattle, rats, mice, cats and other alien species, to the lasting detriment of their vegetation and bird stocks.

During the 20th century, stocks of southern whales, especially humpback whales and the larger rorquals, were hunted excessively despite the attentions of the well-founded International Whaling Commission. During the same period stocks of commercial fish in the subpolar north Atlantic and Pacific oceans were exploited far beyond levels of sustained yield. By contrast, for over 60 years the International Fur Seal Commission provided exemplary management for exploited Pribilov Islands fur seals. For 20 years a small oil industry based on South Georgia elephant seals, which ended when commercial whaling from the island ceased in 1963, was also well managed on sound biological principles.

There is now widespread interest in protecting remnant stocks of polar and subpolar animals, in allowing at least some of the devastated islands to recover, and in preserving what remains of wilderness areas at either end of the world.

7.5.2 *Arctic ownership and management*

The rights of owner-nations to manage and exploit resources on land and in nearshore waters are clearly defined. Where ownership is agreed, responsibility for resource management can be supported by law, and management is generally possible. Where ownership cannot be established, there is neither responsibility nor law; resource management is likely to fail until some equally sound basis for law can be established.

Ownership of arctic lands is nowhere in dispute; every hectare of mainland or island north of 60°N is allocated to one of six nations—Canada, the USA, the USSR, Finland, Denmark, Iceland or Norway. Where doubt over sovereignty has previously been raised, border commissions have secured agreement; even the once-vexed question of Spitsbergen yielded in 1924 to an international treaty that interested parties accepted in preference to continuing dispute.

Each sovereign state has enacted legislation establishing national parks, scientific or scenic reserves and wilderness areas. Usually this has required little more than the extension of existing legislation to cover the special case of Arctic reserves. Wardens are few, but so are visitors to many of the remoter areas. Rules of procedure have been established for minerals exploration, mining and the processing and transport of minerals, and other conservation measures that to one degree or another rationalize exploitation and protect the environment from unnecessary damage. International legislation covers trans-frontier movements of reindeer, caribou and other migratory species. Where conflicts of interest arise (should minerals exploration be permitted in a caribou reserve? see Sheldon, 1988) there are precedents and procedures for resolving them.

Rights to exploit northern marine resources are only slightly less clear. Under the 1982 Convention on the Law of the Sea, 12-mile territorial limits and up-to-200-mile Exclusive Economic Zones (EEZs) have replaced three-mile limits throughout the world, and are as valid in the far north as elsewhere. Subarctic fishing grounds that were once free for all are now controlled by the country in whose EEZ they lie. The 'cod wars' that embroiled Iceland and the UK in the 1970s have been settled by generally-accepted legislation. Though areas of uncertainty and dispute remain, for example in the Barents Sea, settlement is expected within the terms of the 1982 Convention, or at least within the spirit of international cooperation that underlies it (Theutenberg 1984).

The Convention, which is specifically concerned with resource management, endorses scientific research, and encourages international cooperation of the kind that, during the 1970s and early 1980s, investigated the circumpolar status and welfare of polar bears. It covers fisheries, marine mammals and other resources within EEZs, providing for stocks that are shared by neighbouring states. It provides also for management of resources outside the EEZs, both high-seas fish stocks and the resources of the sea bed; the latter it declares to be 'the common heritage of mankind' and places it under the control of an International Sea-bed Authority.

7.5.3 *Management in Antarctica*

Rights over resources in the far south are less clear. Where ownership is undisputed, for example on many of the Southern Ocean islands, rights are clearly assignable and internationally accepted. The French control Iles Kerguelen, Crozet, Amsterdam and St Paul, and lease specified fishing rights to Soviet and other fishermen in circumscribed zones surrounding them. Similarly South Africa controls the Prince Edward Islands, Australia controls Heard and Macquarie islands, New Zealand controls Campbell Island and the Auckland, Bounties, Antipodes and Snares groups. The UK, despite opposition from Argentina, is in *de facto* control of the Falkland Islands, South Georgia and the South Sandwich Islands, and has established an effective fisheries management programme for Falkland Islands waters.

Each controlling nation enacts conservation legislation for its islands, proscribing hunting, settlement, burning, introduction of alien species and other forms of despoilation. Many islands have the status of reserves, with flora, fauna and other natural amenities rigorously protected, at least on paper, through laws enacted by the owning countries.

Sovereignty over continental Antarctica is less clear because ownership is in question. Seven nations (Australia, France, New Zealand, Norway, the UK, Argentina, Chile) claim sectors of the continent; the claims of the last three overlap substantially, and only the first five recognize each others' claims reciprocally. The USA and USSR deny existing sovereignty claims, make no claims of their own, but reserve the right to make claims in the future. Several other nations (Brazil, China, Federal German Republic, German Democratic Republic, India, Japan, Poland and South Africa) have declared interests in Antarctica and established scientific stations there, but lay no claims to territory themselves, and avoid adopting positions on the claims of others.

Laws enacted by claimant nations within the territories claimed, including those concerned with conservation and resource management, could not be regarded as binding by other nations that for any reason rejected the claims. This mattered little while Antarctica remained virtually

Figure 7.2 Polar logistics. Aircraft are used both for exploration and for day-to-day management in both polar regions – effectively, but adding substantially to costs. Photo: Guy Mannering.

empty, as it did up to the end of World War II, but became important in post-war years when scientists and technicians of many nations began to work there. Possible difficulties and conflicts were avoided by the creation, initially under US initiative, of the Antarctic Treaty of 1961, which avoided questions of ownership and provided an alternative basis for conservation legislation.

The Treaty arose from international cooperation during the International Geophysical Year (IGY). The twelve nations most concerned in Antarctic research during the IGY (1957–58) negotiated the original treaty, ratified in 1961, which was and still is open to all nations to join. Since 1961 twenty other nations have signed the Treaty, of which six have been

Figure 7.3 Penguins pursue traditional ways as man explores their Antarctic homeland. Photo: Guy Mannering.

acquired the special consultative status accorded to nations that are active in research. Under Article IV, signatories agree that for the duration of the Treaty (initially 30 years) sovereignty claims will not be pursued or strengthened, i.e. are effectively shelved. New legislation covering the area is agreed among and ratified by the Treaty powers.

The Treaty was not specifically concerned with conservation, but from recommendations discussed at its biannual Consultative Meetings have arisen three conventions that implement conservation, the Convention for the Conservation of Antarctic Flora and Fauna, the Convention for the Conservation of Antarctic Seals, and the Convention on the Conservation of Antarctic Marine Living Resources. The first two relate to the Treaty area, circumscribed by 60°S; the third applies within the much wider ambit of the Antarctic Convergence. By defining sites of special scientific (mostly ecological and geological) interest, and setting upper limits to catches of seals, fish and krill, between them these measures ensure that the Antarctic area will not be despoiled casually, and that its biological resources will not again be exploited without some reference to their capacity for sustained yield. For ecological, legal and political issues arising from the Treaty see Orrego Vicuna (1983), Triggs (1987) and Parsons (1987).

7.6 Summary

Man is a primarily a tropical animal, only recently arrived in polar regions. Aspects of human homeothermy are discussed; for their survival, human populations clearly depend more on skills of dressing, housing and hunting than on physiological adaptation. The ecology of indigenous northern human populations (Inuit, Saami) is outlined. Arctic biological resources have been sufficient to support small, scattered, nomadic populations which lived mainly along the coast or rivers. Southern influences from the 17th century onward, including those of fur traders, whalers, sealers and commercial fisheries, have tended generally to over-exploit natural resources and place indigenous populations at risk. Similarly, stocks of seals and whales in high southern latitudes suffered over-exploitation during the 19th and 20th centuries. Resource management and conservation strategies are currently being attempted for both polar regions. Those for the north, and for many southern cold temperate islands, are based on national legislation drawn up by owning countries. Those for Antarctica are based on the international Antarctic Treaty, which seeks generally-acceptable legislation to cover present and future conservation for the Antarctic area.

REFERENCES

Chapter 1

Andrews, J. (1979) The present ice age: Cenozoic. In John, B.S. (ed.), *The Winters of the World*, David and Charles, Newton Abbot, 173–218.

Armstrong, T.E., Rogers, G. and Rowley, G. (1978) *The Circumpolar North; A Political and Economic Geography of the Arctic and Sub-Arctic.* Methuen, London.

CIA (1978) *Polar Regions Atlas.* Central Intelligence Agency, Washington DC.

Cook, J. (1784) *A Voyage Towards the South Pole and Around the World.* W. Strahan and T. Cadell, London.

Drewry, D.J. (ed.) (1983) *Glaciological and Geophysical Map folio of Antarctica.* Scott Polar Research Institute, Cambridge.

Flint, R.F. (1971) *Glacial and Quaternary Geology.* Wiley, New York.

Govorukha, L.S. and Kruchinin, Yu. A. (1981) *Problems of Physiographic Zoning of Polar Lands.* Amerind, New Delhi.

Huschke, R.E. (1959) *Glossary of Meteorology.* American Meteorological Society, Boston.

Irving, L. (1972) *Arctic Life of Birds and Mammals, Including Man.* Springer, Berlin.

John, B.S. (1979*a*) Planet Earth and its seasons of cold. In John, B.S. (ed.), *The Winters of the World*, David and Charles, Newton Abbot, 9–28.

John, B.S. (1979*b*) The great ice age: Permo-Carboniferous. In John, B.S. (ed.), *The Winters of the World*, David and Charles, Newton Abbot, 154–72.

Kimmins, J.P. and Wein, R.W. (1986) Introduction. In Van Cleve, K., Chapin, F.S., Flanagan, P.W., Viereck, L.A. and Dyrness, C.T. (eds.), *Forest Ecosystems in the Alaskan Taiga: A Synthesis of Structure and Function*, Springer, New York.

Köppen, W. and Geiger, R. (eds.) (1936) *Handbuch der Klimatologie, 1C.* Verlagsbuchhandlung Gebrüder Borntraeger, Berlin.

Kurtén, B. (1972) *The Ice Age*, Hart-Davis, London.

Lamb, H.H. (1972) *Climate: Present, Past and Future.* 1. *Fundamentals and Climate Now.* Methuen, London.

Lamb, H.H. (1977) *Climate: Present, Past and Future.* 2. *Climatic History and the Future.* Methuen, London.

Mill, H.R. (1905) *The Siege of the South Pole.* Alston Rivers, London.

Nilsson, T. (1983) *The Pleistocene.* Reidel, Dordrecht.

Pruitt, W.O. (1978) *Boreal Ecology.* Edward Arnold, London.

Reader's Digest (1985) *Antarctica: Great Stories from the Frozen Continent.* Reader's Digest, London.

Stonehouse, B. (1982) La zonation écologique sous less hautes latitudes australes. *Comité National Français des Recherches Antarctiques* **51**, 531–537.

Sugden, D. (1982) *Arctic and Antarctic: A Modern Geographical Synthesis. Blackwell, Oxford.*

Thornthwaite, C.W. (1948) An approach toward a rational classification of climate. *Geogr. Rev.* **38**, 55–80.

Tukkanen, S. (1980) Climatic parameters and indices in plant geography. *Acta phytogeogr.* Sue. **67**, Uppsala.

Chapter 2

Barrie, L.A. (1986) Arctic air chemistry: an overview. In Stonehouse, B. (ed.), *Arctic Air Pollution*, Cambridge University Press, Cambridge, 5–23.

Benson, C. (1986) Problems of air quality in local arctic and sub-arctic areas, and regional problems of arctic haze. In Stonehouse, B. (ed.), *Arctic Air Pollution,* Cambridge University Press, Cambridge, 69–84.

Bliss, L.C. (1977) *Truelove Lowland, Devon Island, Canada: A High Arctic ecosystem.* University of Alberta Press, Edmonton.

Bryazgin, N.N. (1986) Method of preparing monthly charts of atmospheric precipitation in Antarctica. In Dolgin, I.M. (ed.), *Climate of Antarctica*, Balkema, Rotterdam, 109–116.

Bull, C. (1964) Mean annual surface temperature. Physical characteristics of the Antarctic ice sheet. In Bushnell, V.C. (ed.), *Antarctic Map Folio Series*, Folio 2, American Geographical Society, New York.

Bull, C. (1966) Climatological observations in ice-free areas of Southern Victoria Land, Antarctica. In Rubin, M.J. (ed.), *Studies in Antarctic Meteorology, Antarctic Research Series* **9**, American Geophysical Union, Washington, 177–194.

Bull, C. (1971) Snow accumulation in Antarctica. In Quam, L.O. and Porter, H.D. (eds.), *Research in the Antarctic*, American Association for the Advancement of Science. Washington, 367–421.

Cameron, R.L. and Goldthwait, R.P. (1961) The US-IGY contribution to Antarctic glaciology. *Symposium on Antarctic Glaciology*. International Association for Scientific Hydrology Publications **55**, 7–13.

Collins, N.J. (1977) The growth of mosses in two contrasting communities in the Maritime Antarctic: measurement and prediction of net annual production. In Llano, G.A. (ed.), *Adaptations within Antarctic Ecosystems*, Smithsonian Institution, Washington DC, 921–34.

Courtin, G.M. and Labine, C.L. (1977) Microclimatological studies on Truelove Lowland. In Bliss, L.C. (ed.), *Truelove Lowland, Devon Island, Canada: a High Arctic Ecosystem,* University of Alberta Press, Edmonton, 73–106.

Dalrymple, P.C. 1966. A physical climatology of the Antarctic plateau. In Rubin, M.J. (ed.), *Antarctic Research Series* **9**, *Studies in Antarctic Meteorology*, American Geophysical Union, Washington DC, 195–231.

Farman, J. (1987) What hope for the ozone layer now? *New Scientist*, 12 Nov., 50–54.

Farman, J.C., Gardiner, B.G. and Shanklin, J.D. (1985) Large losses of total ozone in Antarctica reveal seasonal ClO_x/NO_x interactions. *Nature* (*London*) **31**, 207–210.

Friedmann, E.I., McKay, C.P. and Nienow, J.A. (1987) The cryptolithic microbial environment in the Ross Desert of Antarctica: satellite-transmitted continuous nanoclimate data, 1984 to 1986. *Polar Biology* **7**, 273–287.

Gavrilova, M.K. (1963) *Radiation Climate of the Arctic.* Isreal Program for Scientific Translation and National Science Foundation, Washington DC.

Geiger, R. (1965) *The Climate near the Ground.* Harvard University Press, Cambridge, Mass.

Gressitt, J.L. (1962) Ecology and biogeography of land arthropods in Antarctica. *Polar Record* **11** (72), 333–334.

Hardy, R.N. (1972) *Temperature and Animal Life.* Edward Arnold, London.

Jacka, T.H., Christou, L. and Cook, B.J. (1984) A data bank of mean monthly and annual surface temperatures for Antarctica, the Southern Ocean and South Pacific Ocean. *ANARE* Research Notes **22**.

Janetschek, H. (1967) Arthropod ecology of South Victoria Land. In Gressit, J.L. (ed.), *Entomology of Antarctica. Antarctic Research Series* **10**, American Geophysical Union, Washington DC, 205–293.

Larsen, T. (1978) *The World of the Polar Bear*. Hamlyn, London.

Liljequist, G.H. (1956) Energy exchange of an Antarctic snowfield. *Norwegian-British-Swedish Antarctic Expedition*, 1949–52, *Scientific Results* **2** (1).

Loewe, F. (1970) Screen temperature and 10 m temperatures. *J. Glacio.* **9** (56), 263–68.

Loewe, F. (1972) The land of storms. *Weather* **27**(3), 110–121.

MacGregor, K. (1985) Potential for solar energy in northern climates. *Int. J. for Ambient Energy* **6** (3), 115–122.

Madigan, C.T. (1929) Meteorology of the Cape Denison station. *Australasian Antarctic Expedition 1911–14, Scientific Repts*, **B** (4), 1–286.

Matsuda, T. (1968) Ecological study of the moss community and microorganisms in the vicinity of Siowa Station, Antarctica. *Japanese Antarctic Research Expedition Scientific Repts E. Biology*, **29**, 1–58.

Pepper, J. (1954) *The Meteorology of the Falkland Islands and Dependencies 1944-50*. Falkland Islands Government, Stanley.

Prior, M.E. (1962) Some environmental features of Hallet Station, Antarctica, with special reference to soil arthropods. *Pacific Insects* **4**, 681–728.

Rusin, N.P. (1964) *Meteorological and Radiational Regime of Antarctica.* Program for Scientific Translations, Jerusalem.

Schwerdtfeger, W. (1984) *Weather and Climate of the Antarctic.* Elsevier, Amsterdam.

Scott, R.F. (1914) *Scott's Last Expedition.* Smith, Elder, London.

Stonehouse, B. (ed.) (1986) *Arctic Air Pollution.* Cambridge University Press, Cambridge.

Taljaard, J.J. (1972) Synoptic meteorology of the southern hemisphere. In Newton, C.W. (ed.), *Meteorological Monographs* **35**, American Meteorological Society, Boston.

Thompson, D.C. and Macdonald, W.J.P. (1962) Radiation measurements at Scott Base. *N. Z. J.* Geol. and Geophys. **5**, 874–909.

Tilbrook, P.J. (1977) Energy flow through a population of the collembolan *Cryptopygus antarcticus.* In Llano, G.A. (ed.), *Adaptations within Antarctic Ecosystems*, Smithsonian Institution, Washington DC, 935–946.

Vowinckel, E. and Orvig, S. (1970) The climate of the north polar basin. In Orvig, S. (ed.). *Climates of the Polar Regions*, World Survey of Climatology **14**, Elsevier, Amsterdam.

Walton, D.W.H. (1982a) Instruments for measuring biological microclimates for terrestrial habitats in polar and high alpine regions: a review. *Arctic and Alpine Res.* **14**, 275–86.

Walton, D.W.H. (1982*b*) The Signy Island Reference Sites: XIII. Microclimate monitoring 1972–74. *Bull. Br. Antarctic Surv.* **55**, 11–26.

Walton, D.W.H. (1984) The terrestrial environment. In Laws, R.M. (ed.), *Antarctic Ecology* **1**, *Academic Press*, London, 1–60.

WMO (1971) *Climatic Normals (CLINO) for CLIMAT and CLIMAT Ship Stations for the Period 1931–1960.* World Meteorological Organization, Geneva.

Yoshimoto, C.M. and Gressitt, J.L. (1963) Trapping of air-borne insects in the Pacific-Antarctic area, 2. *Pacific Insects* **5** (4), 738–783.

Yoshimoto, C.M., Gressitt, J.L. and Mitchell, C.J. (1962) Trapping of air-borne insects in the Pacific-Antarctic area, 1. *Pacific Insects* **4** (4), 847–858.

Zillman, J.W. (1967) The surface radiation balance in high southern latitudes. In Orvig, S. (ed.), *WMO-SCAR-ICPM Proc. Symp. on Polar Meteorology, WMO Technical Note* **87**, World Meteorological Organization, Geneva.

Chapter 3

Aleksandrova, V.D. (1980) *The Arctic and Antarctic: Their Division into Geobotanical Areas.* Cambridge University Press, Cambridge.

Aleksandrov, M.V. and Simonov, I.M. (1981) Intralandscape zoning of low-lying oases of the eastern Antarctic. In Govorukha, L.S. and Kruchin, Yu. A. (eds.), *Problems of Physiographic Zoning of Polar Lands*, National Science Foundation: Washington DC, 223–242.

Allen, S. E. and Heal, O.W. (1970) Soils of the maritime Antarctic zone. In Holdgate, M.W. (ed.), *Antarctic Ecology* **2**, Academic Press, London, 693–696.
Allen, S.E. and Northover, M.J. (1967) Soil types and nutrients on Signy Island. *Phil. Trans. R. Soc. London B*, **252**, 179–185.
Allison, I., and Keage, P.L. (1986) Recent changes in the glaciers of Heard Island. *Polar Record* **23** (144), 255–271.
Anderson, H.L. and Lent, P.C. (1977) Reproduction and growth of the tundra hare (*Lepus othus*). *J. Mammal.* **58**, 53–57.
Batzli, G.O., White, R.G. and Bunnell, F.L. (1981) Herbivory: a strategy of tundra consumers. In Bliss, L.C., Heal, W.O., and Moore, J.J. (eds.), *Tundra Ecosystems. A Comparative Analysis*, Cambridge University Press, Cambridge, 359–375.
Bee, J.W. and Hall, E.R. (1956) *Mammals of Northern Alaska*. University of Kansas, Kansas City.
Bliss, L.C. (1971) North American and Scandinavian tundras and polar deserts. In Bliss, L.C., Heal, O.W. and Moore, J.J. (eds.), *Tundra Ecosystems. A Comparative Analysis*, 8–24. Cambridge University Press.
Bliss, L.C. (1979) Vascular plant vegetation of the southern circumpolar region in relation to antarctic, alpine and arctic vegetation. *Can. J. Bot.* **57**, 2167–78.
Block, W. (1984) Terrestrial microbiology, invertebrates and ecosystems. In Laws, R.M. (ed.), *Antarctic Ecology* **1**, Academic Press, London, 163–236.
Boyd, W.L. and Boyd, J.W. (1963) Soil micro-organisms of the McMurdo Sound area, Antarctica. *Appl. Microbiol.* **11**, 118–121.
Broady, P.A. (1986) Ecology and taxonomy of the terrestrial algae of the Vestfold Hills. In Pickard, J. (ed.), *Antarctic Oasis: Terrestrial Environments and the History of the Westford Hills*, Academic Press Australia, North Ryde, 165–202.
Broady, P.A., Given, D., Greenfield L. and Thompson, K. (1987) The biota and environment of fumaroles on Mt Melbourne, northern Victoria Land. *Polar Biol.* **7**(2), 97–113.
Cameron, R.E. (1972) Ecology of blue-green algae in antarctic soils. In Desikachary, T.V. (ed.), *First Int. Symp. on Taxonomy and Biology of Blue-Green Algae*, University of Madras, Madras.
Cameron, R.E., King, J. and David, C.N. (1970) Microbiology, ecology 'and microclimatology of soil sites in dry valleys of southern Victoria Land, Antarctica. In Holdgate, M.W. (ed.), *Antarctic Ecology* **2**, Academic Press, London, 702–716.
Campbell, J.C.F. (ed.) (1966) Antarctic soils and soil-forming processes. *Antarctic Research Series* **8**, American Geophysical Union, Washington DC, 161–177.
Campbell, L.B. and Claridge, G.G.C. (1969) A classification of frigic soils—the zonal soils of the Antarctic continent. *Soil Science* **107**, 75–85.
Campbell, L.B. and Claridge, G.G.C. (1987) *Antarctica: Soils, Weathering Processes and Environments*. Elsevier, Amsterdam.
Chambers, M.J.G. (1966a) Investigations of patterned ground at Signy Island, South Orkney Islands. I. Interpretation of mechanical analyses. *Bull. Br. Antarctic Surv.* **9**, 21–40.
Chambers, M.J.G. (1966b) Investigations of patterned ground at Signy Island, South Orkney Islands. II. Temperature regimes in the active layer. *Bull. Br. Antarctic Surv.* **10**, 71–83.
Chambers, M.J.G. (1967) Investigations of patterned ground at Signy Island, South Orkney Islands. III. Miniature patterns, frost heaving and general conclusions. *Bull. Br. Antarctic Surv.* **12**, 1–22.
Chernov, Yu. I. (1985) *The Living Tundra*. Cambridge University Press, Cambridge.
Crawley, M.J. (1983) *Herbivory: The Dynamics of Animal–Plant Interactions*. Blackwell Scientific, Oxford.
Davey, A. (1986) Nitrogen fixation by cyanobacteria in the Vestfold Hills. In Pickard, J. (ed.), *Antarctic Oasis: Terrestrial Environments and the History of the Westford Hills*, Academic Press Australia, North Ryde, 203–220.
Emiliani, C. (1978) The causes of the ice age. *Earth Planet. Sci. Lett.* **37**, 349–352.

Evans, P.R. (1985) Migration. In Campbell, B. and Lack, E. (eds.) *A Dictionary of Birds*, Poyser, Catton, 348–353.

Everett, K.R. (1969) Pedology of the Trinity Peninsula and offshore islands. *Antarctic J. of the United States* **4**, 138.

Evteev, S.A. (1964) At what speed does wind 'erode' stones in Antarctica? *Soviet Antarctic Expedition, Inf. Bull.* **2**, 211.

Fraser, C. (1986) *Beyond the Roaring Forties.* Government Printing Office, Wellington.

French, D.D. and Smith, V.R. (1985) A comparison between northern and southern hemisphere tundras and related ecosystems. *Polar Biol.* **5**, 5–21.

Gimingham, C.H. and Smith, R.I.L. (1970) Bryophyte and lichen communities in the Maritime Antarctic. In Holdgate, M.W. (ed.), *Antarctic Ecology* **2**, Academic Press, London, 752–785.

Glazovskaya, M.A. (1958) Weathering and primary soil formations in Antarctica. *Scientific Papers of the Institute, Moscow University, Faculty of Geography* **1**, 63–76 (In Russian).

Goldthwait, R.P., Dreimanis, A., Forsyth, J.L., Carrow, P.F. and White, G.W. (1965) Pleistocene deposits of the Erie Lobe. In Wright, H.E. and Frey, D.G. (eds.), *The Quaternary of the United States*, Princeton University Press, Princeton, 85–97.

Greene, S.W. (1964) Plants of the land. In Priestley, R., Adie, R.J. and Robin, G. de Q. (eds.), *Antarctic Research*, Butterworth, London.

Greene, S.W. and Walton, D.W.H. (1975) An annotated check list of the sub-antarctic and antarctic vascular flora. *Polar Record* **17** (110), 473–484.

Haber, E. (1986) Flora of the circumpolar Arctic. In Sage, B. (ed.), *The Arctic and its Wildlife*, Croom Helm, Beckenham.

Hays, J.D., Imbrie, J. and Shackleton, N.J. (1976) Variations in the earth's orbit: pacemaker of the ice ages. *Science* **194**, 1121–1132.

Holdgate, M.W. (1964) Terrestrial ecology in the maritime Antarctic. In Carrick, R., Holdgate, M. and Prévost, J. (eds.), *Biologie antarctique*, Hermann, Paris, 181–194.

Holdgate, M.W. (1977) Terrestrial ecosystems in the Antarctic. *Phil. Trans. R. Soc. London B* **279**, 5–25.

Holdgate, M.W., Allen, S.E. and Chambers, M.J.G. (1967) A preliminary investigation of the soils of Signy Island. *Bull. Br. Antarctic Surv.* **12**, 53–71.

Hutchinson, G.E. (1950) Biogeochemistry of vertebrate excretion. *Bull. Amer. Mus. Nat. Hist.* **96**.

Imbrie, J. and Imbrie, K.P. (1979) *Ice Ages: Solving the Mystery.* Enslow Publishers, Shore Hills.

Irving, L. (1972) *Arctic Life of Birds and Mammals, Including Man.* Springer, Berlin.

Janetschek, H. (1963) On the terrestrial fauna of the Ross Sea area, Antarctica. *Pacific Insects* **5**, 305–311.

Janetschek, H. (1967) Arthropod ecology of South Victoria Land. *Antarctic Res. Ser.* **10**, 205–293.

Jones, N.V. (1963) The sheathbill, *Chionis alba* (Gmelin), at Signy Island, South Orkney Islands. *Bri. Antarctic Surv. Bull.* **2**, 53–71.

Korotkevich, E.S. (1971) Quaternary marine deposits and terraces in Antarctica. *Inf. Bull. of the Soviet Antarctic Expeditions* **82**, 185–190.

Kubiena, W.L. (1953) *The Soils of Europe.* Murby, London.

Kurtén, B. (1968) *Pleistocene Mammals of Europe.* Aldine, Chicago.

Lamb, I.M. (1970) Antarctic terrestrial plants and their ecology. In Holdgate, M.W. (ed.) *Antarctic Ecology* **2**, Academic Press, London, 733–751.

Leader-Williams, N. (1985) The sub-antarctic islands—introduced mammals. In Bonner, W.N. and Walton, D.W.H. (eds.), *Antarctica*, Pergamon, Oxford, 318–328.

Lesel, R. and Derenne, P. (1975) Introducing animals to Iles Kerguelen. *Polar Record* **17**, 485–494.

Lindsey, C.C. (1981) Arctic refugia and the evolution of arctic biota. In Scudder, G.G. and Reveal, J.L. (eds.), *Evolution Today*, Proc. Second Int. Congr. of Systematic and

Evolutionary Biology, Hunt Institute for Botanical Documentation, Carnegie-Mellon University, Pittsburgh, 7–10.

Llano, G.A. (1965) The flora of Antarctica. In Hatherton, T. (ed.), *Antarctica: A New Zealand Antarctic Society Survey*, Methuen, London.

MacNamara, E.E. (1969) Active layer development and soil moisture dynamics in Enderby Land, East Antarctica. *Soil Science* **108**, 345–349.

McCraw, A. (1960) Soils of the Ross Dependency, Antarctica. *Proc. N. Zeal. Soil Science Soc.* **4**.

Minns, C.K. (1977) Limnology of some lakes on Truelove Island. In Bliss, L.C. (ed.), *Truelove Lowland, Devon Island, Canada: a High* Arctic ecosystem. University of Alberta Press, Edmonton, 566–585.

Myrcha, A., Pietr, S.J. and Tatur, A. (1985) The role of pygoscelid penguin rookeries in nutrient cycles at Admiralty Bay, King George Island. In Siegfried, W.R., Condy, P.R. and Laws, R.M. (eds.), *Antarctic Nutrient Cycles and Food Webs*, Springer, Berlin.

Nichols, R.L. (1966) Geomorphic features in the McMurdo Sound area. *Geotimes* **11** (4), 19–22.

Nichols, R.L. and Ball, D.G. (1964) Soil temperatures, Marble Point, McMurdo Sound, Antarctica. *J. Glaciol.* **5** (39), 357–359.

Orchard, V.A. and Corderoy, D.M. (1983) Influence of environmental factors on the decomposition of penguin guano in Antarctica. *Polar Biol.* **1**, 199–204.

Pattie, D.L. (1972) Preliminary bioenergetic and population level studies in High Arctic birds. In Bliss, L.C. (ed.) *Devon Island IBP Project, High Arctic ecosystems.* Edmonton, University of Alberta Press, 281–292.

Pattie, D.L. 1977. Population levels and bioenergetics of Arctic birds on Truelove Lowland. In Bliss, L.C. *Truelove Lowland, Devon Island, Canada: A High Arctic Ecosystem*, University of Alberta Press, Edmonton, 413–433.

Pickard, J. (1986*a*) Antarctic oases, Davis Station and the Vestfold Hills. In Pickard, J. (ed.), *Antarctic Oasis: Terrestrial environments and the History of the Westford Hills*, Academic Press Australia, North Ryde, 1–19.

Pickard, J. (1986*b*) Spatial relations of the vegetation of the Vestfold Hills. In Pickard, J. (ed.) *Antarctic Oasis: Terrestrial Environments and the History of the Westford Hills*, Academic Press Australia, North Ryde, 275–308.

Pickard, J. (ed.) (1986*c*) *Antarctic Oasis: Terrestrial Environments and the History of the Westford Hills*, Academic Press Australia, North Ryde.

Pickard, J. and Seppelt, R.D. (1984) Phytogeography of Antarctica. *J. Biogeogr.* **11**, 83–102.

Péwé, T.L. (1960) Multiple glaciation in the McMurdo Sound region, Antarctica—a progress report. *J. Geol.* (*Chicago*) **68**, 498–514.

Richardson, D.H.S. and Finegan, E.J. (1977) Studies on the lichens of Truelove Lowland. In Bliss, L.C. (ed.) *Truelove Lowland, Devon Island, Canada: A High Arctic Ecosystem*, University of Alberta Press, Edmonton, 245–262.

Ricklefs, R.E. (1980) Geographical variations in clutch size among passerine birds: Ashmole's hypothesis. *Auk* **97**, 38–49.

Ròzycki, S.Z. (1961) Changements Pleistocènes de l'extension de l'inlandis en Antarctide orientale d'après l'étude des anciennes plages elevées de l'Oasis Bunger, Queen Mary's Land. *Biuletyn Periglacjalny* **10**, 257–283.

Rudolph, E.D. (1966) Terrestrial vegetation of Antarctica: past and present studies. Antarctic soils and soil forming processes. *Antarctic Res. Ser.* **8**, 109–124.

Ryan, J.K. (1977) Synthesis of energy flows and population dynamics of Truelove Lowland invertebrates. In Bliss, L.C. (ed.), *Truelove Lowland, Devon Island, Canada: A High Arctic Ecosystem*, University of Alberta Press, Edmonton, 325–346.

Sage, B. (1986*a*) Breeding birds. In Sage, B. (ed.), *The Arctic and its Wildlife*, Croom Helm, London.

Sage, B. (1986*b*) Terrestrial mammals. In Sage, B. (ed.), *The Arctic and its Wildlife*. Croom Helm, London.
Seppelt, R.D. (1986*a*) Bryophytes of the Westford Hills. In Pickard, J. (ed.), *Antarctic Oasis: Terrestrial Environments and the History of the Westford Hills*, Academic Press Australia, North Ryde, 221–245.
Seppelt, R.D. (1986*b*) Lichens of the Westford Hills. In Pickard, J. (ed.), *Antarctic Oasis: Terrestrial Environments and the History of the Westford Hills*, Academic Press Australia, North Ryde, 247–274.
Shumskiy, P.A. (1957) Glaciological and geomorphological reconnaissance in the Antarctic in 1956. *J. Glaciol.* **3**, 56–61.
Smith, R.I.L. (1984) Terrestrial plant biology of the sub-Antarctic and Antarctic. In Laws, R.M. (ed.), *Antarctic Ecology* **2**, Academic Press, London, 61–162.
Smith, R.I.L. and Poncet, S. (1985) New southernmost record for Antarctic flowering plants. *Polar Record* **22** (139), 425–427.
Solopov, A.V. (1969) *Oases in Antarctica.* National Science Foundation, Washington DC.
Stonehouse, B. (1971) *Animals of the Arctic: The Ecology of the Far North.* Ward Lock, London.
Stonehouse, B. (1972) *Animals of the Antarctic: The Ecology oj 'he Far South.* Peter Lowe, London.
Stonehouse, B. (1982) La zonation écologique sous les hautes latitudes australes. In Jouventin, P., Massé, L., and Tréhen, P. (eds.), *Colloque sur les ecosystèmes subantarctiques*, Comité national français des recherches antarctiques Paris, 531–536.
Stonehouse, B. (1985) Sheathbill. In Campbell, B. and Lack, E. (eds.), *A Dictionary of Birds*, Poyser, Calton, 533.
Syroechkovskiy, E.E. (1959) The role of animals in primary soil formation in the conditions of the circumpolar regions of the earth (exemplified by the Antarctic). *Zool Zh. Ukraine* **38**, 1770–1775 (in Russian).
Tedrow, J.C.F. (1977) *Soils of the Polar Landscapes.* Rutgers University Press, New Brunswick.
Tedrow, J.C.F. and Brown, J. (1967) Soils of Arctic Alaska. In Wright, H.E. and Osburn, W.H. (eds.), *Arctic and Alpine Environments*, Indiana University Press, Bloomington, 283–93.
Tedrow, J.C.F. and Ugolini, F.C. (1966) Antarctic soils. In Campbell, J.C.F. (ed.), *Antarctic Soils and Soil-Forming Processes, Antarctic Research Series* **8**, American Geophysical Union, Washington DC, 161–177.
Tikhomirov, B.A. (1961) The treelessness of the tundra. *Polar Record* **11** (70), 24–30.
Tryon, P.R. and MacLean, S.F. (1980) Use of space by Lapland longspurs breeding in Arctic Alaska. *Auk* **97**, 509–520.
Ugolini, F.C. (1966) Soil investigations in the Lower Wright Valley, Antarctica. *Proc. Int. Conf. on Permafrost*, National Academy of Sciences, National Research Council Publication **1287**, Washington DC, 55–61.
Ugolini, F.C. (1967) Soils of Mount Erebus, Antarctica. *N. Z. J. Geol. and Geophys.* **10** (2), 431–432.
Ugolini, F.C. (1970) Antarctic soils and their ecology. In Holdgate, M.W. (ed.), *Antarctic Ecology* **2**, Academic Press, London, 702–716.
Ugolini, F.C. (1972) Ornithogenic soils of Antarctica. In Llano, G.A. (ed.), *Antarctic Research Series* **20**, American Geophysical Union, Washington DC, 181–193.
Ugolini, F.C. (1977). The protoranker soils and the evolution of an ecosystem at Kar Plateau, Antarctica. In Llano, G.A. (ed.) *Adaptations Within Antarctic Ecosystems*, Smithsonian Institution, Washington DC, 935–946.
Ugolini, F.C. (1986) Processes and rates of weathering in cold and polar desert environments. In Colman, S.M. and Dethier, D.P. (eds.), *Rates of Chemical Weathering of Rocks and Minerals*, Academic Press, New York.
Walker, B.D. and Peters, T.W. (1977) Soils of Truelove Lowland and plateau. In Bliss, L.C.

(ed.), *Truelove Lowland, Devon Island, Canada: A High Arctic Ecosystem*, University of Alberta Press, Edmonton, 31–62.
Walton, D.W.H. (1985) A preliminary study of the action of crustose lichens on rock surfaces in Antarctica. In Siegfried, W.R., Condy, P.R. and Laws, R.M. (eds.), *Antarctic Nutrient Cycles and Food Webs*, Springer, Berlin.
Washburn, A.L. (1979) *Geocryology: a Survey of Periglacial Processes and Environments.* Edward Arnold, London.
Watson, A. (1957) The behaviour, breeding and food-ecology of the snowy owl *Nyctea scandiaca. Ibis* **99**, 419–462.
Watson, G.E., Angle, J.P. and Harper, P.C. (1975) *Birds of the Antarctic and Sub-Antarctic.* American Geophysical Union, Washington DC.
Weller, M.W. (1975) Ecology and behaviour of the South Georgia pintail *Anas g. georgica. Ibis* **117**, 227–231.
Weller, M.W. and Howard, R.L. (1972) Breeding of speckled teal *Anas flavirostris* on South Georgia. *Bull. Br. Antarctic Surv.* **30**, 65–68.
Wellman, H.W. and Wilson, A.T. (1965) Salt weathering, a neglected geological erosive agent in coastal and arid environments. *Nature* (London) **205**, 1097–1098.
West, G.C. (1972) Seasonal differences in resting metabolic rate of Alaskan ptarmigan. *Compar. Biochem. Physiol.* **42**, 867–876.
Wiegolaski, F.E. (1972) Vegetation types and primary production in tundra. In Wiegolaski, F.E. and Rosswall, T. (eds.), *IBP Tundra Biome: Proc. IV Int. Meeting on the Biological Productivity of Tundra*, Leningrad, USSR, October 1971, Tundra Biome Steering Committee, Stockholm, 1–34.
Wisniewski, E. (1983) Bunger Oasis: the largest ice-free area in the Antarctic. *Terra* **95**, 178–187.
Wrenn, J.H. and Webb, P.N. (1982) Physiographic analysis and interpretation of the Ferrar Glacier-Victoria Valley area, Antarctica. In Craddock, C. (ed.), *Antarctic Geoscience*, University of Wisconsin Press, Madison, 1091–1099.

Chapter 4

Adamson, D.A. and Pickard, J. (1986) Cainozoic history of the Westford Hills. In Pickard, J. (ed.), *Antarctic Oasis: Terrestrial Environments and History of the Westford Hills*, Academic Press Australia, North Ryde, 63–97.
Akiyama, M. (1979) Some ecological and taxonomic observations on the coloured snow algae found in Rumpa and Skarvsnes, Antarctica. In Matsuda, T. and Hoshiai, T. (eds.), *Proc. Symp. on Terrestrial Ecosystems in the Syowa Station Areas*, National Institute of Polar Research, Tokyo, 27–34.
Benninghoff W.S. and Benninghoff A.S. (1985) Wind transport of electrostatically charged particles and minute organisms in Antarctica. In Siegfried, W.R., Condy, P.R. and Laws, R.M. (eds.), *Antarctic Nutrient Cycles and Food Webs*, Springer, Berlin, 592–596.
Craig, P.C. and McCart, P.J. (1975) Classification of stream types in Beaufort Sea drainages between Prudhoe Bay and the Mackenzie Delta, NWT, Canada. *Arctic and Alpine Res.* **7**, 183–198.
Dickman, M. and Ouelett, M. (1987) Limnology of Garrow Lake, NWT, Canada. *Polar Record* **23** (146), 531–549.
Ellis-Evans, J.C. (1981*a*) Freshwater microbiology in Antarctica. I, Microbial numbers and activity in oligotrophic Moss Lake. *Bull. Br. Antarctic Surv.* **54**, 85–104.
Ellis-Evans, J.C. (1981*b*) Freshwater microbiology in Antarctica. II, Microbial numbers and activity in nutrient-enriched Heywood Lake. *Bull. Br. Antarctic Surv.* **54**, 105–121.
Ellis-Evans, J.C. (1982) Seasonal microbial activity in Antarctic freshwater lake sediments. *Polar Biol.* **1**, 129–140.
Ellis-Evans, J.C. (1985*a*) Decomposition processes in maritime Antarctic lakes. In Siegfried, W.R., Condy, P.R. and Laws, R.M. (eds.), *Antarctic Nutrient Cycles and Food Webs*, Springer, Berlin, 253–260.

Ellis-Evans, J.C. (1985*b*) Fungi from maritime freshwater environments. *Bull. Br. Antarctic Surv.* **68**, 37–45.

Ellis-Evans, J.C. and Wynne-Williams, D.D. (1985) The interaction of soil and lake microflora at Signy Island. In Siegfried, W.R., Condy, P.R. and Laws, R.M. (eds.) *Antarctic Nutrient Cycles and Food Webs*. Springer, Berlin, 662–668.

Fogg, G.E. (1967) Observations on the snow algae of the South Orkney Islands. *Phil. Trans. R. Soc. London B* **252**, 279–287.

Goldman, C.R. (1970) Antarctic freshwater ecosystems. In Holdgate, M.W. (ed.), *Antarctic Ecology* **2**, Academic Press, London, 609–631.

Hambrey, M.J. (1984) Sudden drainage of ice-dammed lakes in Spitsbergen. *Polar Record* **22** (137) 189–194.

Harris, H.J.H, Cartwright, K. and Torii, T. (1979) Dynamic chemical equilibrium in a polar desert pond: a sensitive index of meteorological cycles. *Science* **204**, 301–303.

Hawes, I. (1985) Factors controlling phytoplankton populations in Maritime Antarctic lakes. In Siegfried, W.R., Condy, P.R. and Laws, R.M. (eds.), *Antarctic Nutrient Cycles and Food Webs*, Springer, Berlin, 245–252.

Hansen, K. (1967) The general limnology of arctic lakes as illustrated by examples from Greenland. *Meddelelser om Grønland* **77**.

Hattersley-Smith, G. and others (1970) Density stratified lakes in northern Ellesmere Island. *Nature* (*London*) **225**, 55–56.

Hendy, C.H., Wilson, A.T., Popplewell, K.B., and House, J.E. (1977) Dating of geochemical events in Lake Bonney, Antarctica, and their relation to glacial and climatic change. *N. Z. J. Geol. Geophys.* **20**, 1103–1122.

Heywood, R.B. (1977) Antarctic freshwater ecosystems: review and synthesis. In Llano, G.A. (ed.), *Adaptations within Antarctic Ecosystems*, Smithsonian Institution, Washington DC.

Heywood, R.B. (1984) Antarctic inland waters. In Laws, R.M. (ed.), *Antarctic Ecology* **1**, Academic Press, London, 279–344.

Heywood, R.B. (1987) Limnological studies in the Antarctic Peninsula region. In El-Sayed, S.Z. and Tomo, A.P. (eds.) *Antarctic Aquatic Biology*, SCAR (BIOMASS 7), Cambridge, 157–173.

Heywood, R.B., Dartnall, H.J.G. and Priddle, J. (1980) Characteristics and classification of the lakes of Signy Island, South Orkney Islands, Antarctica. *Freshwater Biology* **10**, 47–59.

Hobbie, J.E. (1962) Limnological cycles and primary productivity of two lakes in the Alaskan Arctic. Ph.D. thesis, Indiana University.

Hobbie, J.E. (1964) Carbon-14 measurements of primary production in two arctic Alaskan lakes. *Verh. Int. Verein. theor. angew. Limnol.* **15**, 360–364.

Hoham, R.W. (1975) Optimum temperatures and temperature ranges for growth of snow algae. *Arctic and Alpine Res.* **7**, 13–24.

Hoham, R.W. (1980) Unicellular chlorophytes—snow algae. In Cox, E.R. (ed.), *Phytoflagellates*, Elsevier, New York, 61–84.

Holeton, G.F. (1973) Respiration of arctic char (*Salvelinus alpinus*) from a high arctic lake. *J. Fisheries Res. Bd Canada* **30**, 717–723.

Holmgren, S. (1968) Phytoplankton production in a lake north of the Arctic Circle. Licentiate thesis, University of Uppsala.

Kalff, J. (1970) Arctic lake ecosystems. In Holdgate, M.W. (ed.) *Antarctic Ecology* **2**, Academic Press, London, 651–663.

Kalff, J., Welch, H.E., and Holmgren, S.K. (1972) Pigment cycles in two high arctic Canadian lakes. *Verh. Int. Verein. Limnol.* **18** (1), 250–256.

Kerry, K.R., Grace, D.R., Williams, R. and Burton, H.R. (1977) Studies on some saline lakes of the Vestfold Hills, Antarctica. In Llano, G. (ed.), *Adaptations within Antarctic Ecosystems*, Smithsonian Institution. Washington. 839–858.

Kieffer, L.A. and Copes, C.D. (1987) Limnological studieson Deception Island, South

Shetland Islands: morphology and origin of lentic environments. In El-Sayed, S.Z. and Tomo, A.P. (eds.) *Antarctic Aquatic Biology*, SCAR (Biomass 7), Cambridge, 1207–1217.

Kol, E. and Flint, E.A. (1968) Algae in green ice from the Balleny Islands, Antarctica. *N. Z. J. Botany* **6**, 249–261.

Lasenby, D.C. and Langford, R.R. (1972) Growth, life history respiration of *Mysis relicta* in an arctic and temperate lake. *J. Fisheries Res. Bd Canada* **29**, 1701–1708.

Lasenby, D.C. and Langford, R.R. (1973) Feeding and assimilation of *Mysis relicta. Limnol. and Oceanogr.* **18**, 280–285.

Light, J.J., Ellis-Evans, J.C. and Priddle, J. (1981) Phytoplankton ecology in an Antarctic lake. *Freshwater Biol.* **11**, 11–26.

Mackay, D.K. and Loken, O.H. (1974) Arctic hydrology. In Ives, J.D. and Barry, R.G. (eds.), *Arctic and Alpine Environments*, Methuen, London, 111–132.

Meyer-Rochow, V.B. (1979) Kleinstlebewesen in extremen Biotopenzum 'Tuempeln' in die Antarktis. *Mikrokosmos* **67**, 34–38.

Minns, C.K. (1977) Limnology of some lakes on Truelove Lowland. In Bliss, L.C. (ed.), *Truelove Lowland, Canada: A High Arctic Ecosystem*, University of Alberta Press, Edmonton, 569–585.

Morgan, K. and Kalff, J. (1972) Bacterial dynamics in two high-arctic lakes. *Freshwater Biol.* **2**, 217–228.

Nelson, K.H. and Thompson, T.G. (1954) Deposition of salts from sea water by frigid concentration. *J. Marine Res.* **13**, 166–182.

Nurminen, M. (1973) Enchytraeidae (Oligochaeta) from the arctic archipelago of Canada. *Annal. Zool. Fennici* **10**, 403–411.

Paggi, J.C. (1987) Limnological studies in the Potter Peninsula, 25 de Mayo Island, South Shetland Islands: biomass and spatial distribution of zooplankton. In El-Sayed, S.Z. and Tomo, A.P. (eds.) *Antarctic Aquatic Biology*. SCAR (Biomass 7), Cambridge, 175–191.

Parker, B.C. and Simmons, G.M. (1985) Paucity of nutrient cycling and absence of food chains in the unique lakes of southern Victoria Land. In Siegfried, W.R., Condy, P.R. and Laws, R.M. (eds.) *Antarctic Nutrient Cycles and Food Webs*, Springer, Berlin, 238–244.

Parker, B.C., Keiskell, L.E., Thompson, W.J. and Zeller, E.J. (1978) Non-biogenic fixed nitrogen in Antarctica and some ecological implications. *Nature* (*London*) **227**, 651–652.

Priddle, J. and Heywood, R.B. (1980) Evolution of Antarctic lake ecosystems. *Biol. J. Linn. Soc.* **14**, 51–66.

Priddle, J., Hawes, I. and Ellis-Evans, J.C. (1986) Antarctic aquatic ecosystems as habitats for phytoplankton. *Biol. Rev.* **61**, 199–238.

Rigler, F.H. (1978) Limnology in the high arctic; a case study of Char Lake. *Verh. Int. Verein, Limnol*, **20**, 127–140.

Roff, J.C. (1972) Aspects of the reproductive biology of the planktonic copepod *Limnocalanus macrurus* Sars, 1863. *Crustaceana* **22**, 155–160.

Roff, J.C. (1973) Oxygen consumption of *Limnocalanus macrurus* Sars (Calanoida, Copepoda) in relation to environmental conditions. *Can. J. Zool.* **51**, 877–885.

Roff, J.C. and Carter, J.C.H. (1972) Life cycle and seasonal abundance of the copepod *Limnocalanus macrurus* Sars in a high arctic lake. *Limnol. and Oceanogr.* **17** (3), 363–370.

Thompson, T.G. and Nelson, K.H. (1956) Concentration of brines and deposition of salts from sea water under frigid conditions. *Amer. J. Sci.* **254**, 227–238.

Vincent, W.F. (1988) *Microbial Ecosystems of Antarctica*. Cambridge University Press, Cambridge.

Welch, H.E. (1973) Emergence of Chironomidae (Diptera) from Char Lake, Resolute, Northwest Territories. *Can. J. Zool.* **51**, 1113–1123.

Wharton, R.A., Vinyard, W.C., Parker, B.C., Simmons, G.M. and Seaburg, G. (1981) Algae in cryoconite holes on Canada Glacier in southern Victoria Land, Antarctica. *Phycologia* **20**, 208–211.

Wright, S.W. and Burton, H.R. (1981) Biology of Antarctic saline lakes. *Hydrobiologia* **82**, 319–338.

Chapter 5

Alexander, V. (1974) Primary productivity regimes of the nearshore Beaufort Sea with reference to the potential role of ice biota. In Reed, J.C., and Sater, J.E. (eds.), *The Coast and Shelf of the Beaufort Sea*, Arctic Institute of North America, Arlington, 609–632.

Bakayev, V.G., and others (1966) *Atlas Antarktiki.* Ministry of Geology of the USSR, Moscow.

Balech, E., and others (1968) Primary productivity and benthic marine algae of the Antarctic and subantarctic. *Antarctic Map Folio Series* **10**, American Geophysical Union, Washington, DC.

Barry, R.G. (1987) Aspects of the meteorology of the seasonal sea ice zone. In Untersteiner, N. (ed.), *The Geophysics of Sea Ice*, Plenum, New York, 993–1020.

Blacker, R.W. (1965) Recent changes in the benthos of the West Spitsbergen fishing grounds. *ICNAF Spec. Publ.* **6**, 791–794.

Bonner, W.N. (1985) Birds and mammals—Antarctic seals. In Bonner, W.N., and Walton, D.W.H. (eds.), *Antarctica*, Pergamon, Oxford, 201–222.

Boyd, C.M., Heyraud, M. and Boyd, C.N. (1984) Feeding of the Antarctic krill *Euphausia superba*. *J. Crustacean Biol.* **4**(1), 123–141.

Brodsky, K.A. (1956) Life in the water column of the Polar Basin. *Priroda* **5** (In Russian).

Brodsky, K.A. and Pavshtiks, Y.A. (1976) Plankton of the central part of the Arctic basin (based on collections of the North Pole drifting stations. *Polar Geogr.* **1**(2), 143–161.

Brownell, R.L., Best, P.B. and Prescott, J.H. (eds.) (1986) *Right Whales: Past and Present Status.* International Whaling Commission, Cambridge.

Bursa, A.S. (1961) The annual oceanographic cycle at Igloolik, in the Canadian Arctic. II. The phytoplankton. *J. Fisheries Res. Bd Canada* **18** (4), 563–615.

Cairns, A.A. (1967) The zooplankton of Tanquary Fiord, Ellesmere Island, with special reference to Calanoid copepods. *J. Fisheries Res. Bd Canada* **24** (3), 555–568.

CIA (1978) *Polar Regions Atlas.* Central Intelligence Agency, Washington.

Clasby, R.C., Alexander, V. and Horner, R. (1976) Primary productivity of sea-ice algae. In Hood, D.W. and Burrell, D.C. (eds.), *Assessment of the Arctic Marine Environment. Selected Topics, Occas. Publ.* **4**, Institute of Marine Sciences, University of Alaska, Fairbooks, 289–304.

Coachman, L.K. and Aagaard, K. (1974) Physical oceanography of Arctic and Subarctic seas. In Herman, Y. (ed.), *Marine Geology and Oceanography of the Arctic Seas*, Springer, New York, 1–72.

Croxall, J.P. (1984) Seabirds. In Laws, R.M. (ed.), *Antarctic Ecology*, Academic Press, London, 533–616.

Croxall, J.P. (ed.) (1987) *Seabirds: Feeding Ecology and Role in Marine Ecosystems*, Cambridge University Press, Cambridge.

Croxall, J.P. and Lishman, G.S. (1987) The food and feeding ecology of penguins. In Croxall, J.P. (ed.) *Seabirds: Feeding Ecology and Role in Marine Ecosystems*, Cambridge University Press, Cambridge, 101–133.

Croxall, J.P., Evans, P.G.H. and Schreiber, R.W. (1984) *Status and Conservation of the World's Seabirds*, International Council for the Protection of Birds, Cambridge.

Dinofrio, E.O. (1987) Planktonic copepods of waters near the South Orkney Islands. In El-Sayed, S.Z. and Tomo, A.P. (ed.), *Antarctic Aquatic Biology*, SCAR (Biomass 7), Cambridge, 53–65.

Dunbar, M.J. (1968) *Ecological Development in Polar Regions: A Study in Evolution.* Prentice-Hall, Englewood Cliffs.

Dunbar, M.J. (1982) Arctic marine ecosystems. In Rey, L. and Stonehouse, B. (eds.), *The Arctic Ocean*, MacMillan, London, 233–261.

Dunbar, M.J. and Harding, G. (1968) Arctic Ocean water masses and plankton: a reappraisal. In Sater, J.E. (ed.), *Arctic Drifting Stations*, Arctic Institute of North America, Arlington, 315–326.

Eicken, H., Grenfell, T.C. and Stonehouse, B. (1988) Sea ice conditions during an early spring voyage in the eastern Weddell Sea, Antarctica. *Polar Record* **24**(148), 49–54.

El-Sayed, S.Z. (1971) Observations on plankton bloom in the Weddell Sea. In Llano, G.A. and Wallen, I.E. (eds.), *Biology of the Antarctic Seas* IV, American Geophysical Union, Washington DC.

El-Sayed, S.Z. (1985) Plankton of the Antarctic seas. In Bonnor, W.N. and Walton, D.W.H. (eds.), *Key Environments: Antarctica.* Pergamon, Oxford, 135–153.

El-Sayed, S.Z. (1987) Biological productivity of Antarctic waters: present paradoxes and emerging paradigms. In El-Sayed, S.Z. and Tomo, A.P. (eds.), *Antarctic Aquatic Biology.* SCAR (Biomass 7), Cambridge, 1–21.

El-Sayed, S.Z. and Taguchi, S. (1981) Primary production and standing crop of phytoplankton along the ice-edge in the Weddell Sea. *Deep-Sea Research* **28**, 1017–1032.

Everson, I. (1977) *The Living Resources of the Southern Ocean.* Food and Agriculture Organization of the United Nations, Rome.

Ferrigno, J.G. and Gould, W.G. (1987) Substantial changes in the coastline of Antarctica revealed by satellite imagery. *Polar Record* **23**(146), 577–583.

Fogg, G.E. (1977) Aquatic primary production in the Antarctic. *Proc. R. Soc. London B* **279**, 27–38.

Foster, T.D. (1984) The marine environment. In Laws, R.M. (ed.), *Antarctic Ecology* **2**, Academic Press, London, 345–371.

Gambell, R. (1985) Birds and mammals—Antarctic whales. In Bonner, W.N. and Walton, D.W.H. (eds.), *Antarctica*, Fergamon, Oxford, 223–241.

Gierloff-Emden, H.G. (1982) *Das Eis des Meeres.* de Gruyter, Berlin.

Gieskes, W.W.C., Veth, C., Woehrmann and Graefe, M. (1987) Secchi disc visibility world record shattered. *Eos* **89**(9), 23.

Grainger, E.H. (1979) Primary production in Frobisher Bay, Arctic Canada. In Dunbar, M.J. (ed.), *Marine Production Mechanisms*, Cambridge University Press, Cambridge, 9–30.

Gulland, J.A. (1974) *The Management of Marine Fisheries.* University of Washington Press, Seattle.

Hardy, A.C. and Gunther, E.R. (1936) The plankton of the South Georgia whaling grounds and adjacent waters, 1926–27. *Discovery Reports* **11**, 1–456.

Hedgpeth, J.W. (1971) Perspectives of benthic ecology in Antarctica. In Quam, L.O. (ed.), *Research in the Antarctic*, American Association for the Advancement of Science, Washington DC.

Hempel, G. (1985) On the biology of polar seas, particularly the Southern Ocean. In Gray, J.S. and Christiansen, M.E. (eds.), *Marine Biology of Polar Seas and Effects of Stress on Marine Organisms*, Wiley, New York, 3–33.

Hopkins, T.L. (1971) Zooplankton standing crop in the Pacific sector of the Antarctic. In Llano, G.A. and Wallen, I.E. *Biology of the Antarctic Sea* IV, American Geophysical Union, Washington DC, 347–362.

Jeffries, M.O. (1987) The growth, structure and disintegration of Arctic ice shelves. *Polar Record* **23**(147), 631–649.

Jouventin, P. and Stonehouse, B. (1985) Biological survey of Iles de Croy, Iles Kerguelen, 1984. *Polar Record* **22**(141), 688–691.

King, J. (1983) *Seals of the World.* British Museum (Natural History), London.

Knox, G.A. (1970) Antarctic marine ecosystems. In Holdgate, M.W. (ed.), *Antarctic Ecology* **1**, Academic Press, London, 70–96.

Knox, G.A. and Lowry, J.K. (1977) A comparison between the benthos of the Southern Ocean and the North Polar Ocean with special reference to the Amphipoda and Polychaeta. In Dunbar, M.J. (ed.) *Polar Oceans*, Arctic Institute of North America, Calgary, 423–462.

Marr, J. (1962) The natural history and geography of the Antarctic krill (*Euphausia superba* Dana). *Discovery Repts.* **32**, 33–464.
Marshall, N.B. (1982) Glimpses into deep-sea biology. In Rey, L. and Stonehouse, B. (eds.), *The Arctic Ocean,* 263–271.
McClatchie, S. and Boyd, C.M. (1963) A morphological study of sieve efficiency and mandibular surfaces in the Antarctic krill, *Euphausia superba. J. Fisheries and Aquatic Sci.* **40**, 955–967.
Norderhaug, M., Brun, E. and Mo/llen, C.U. (1977) Barentshavets sjöfuglressurser. *Norsk Polarinstitutt Meddelellser* **104**, 1–119.
Polar Record (1987) The world's clearest sea water off Antarctica. *Polar Record* **23**(147), 737.
Polar Record (1988) Bay of Whales disappears. *Polar Record* **24**(148), 75.
Prince, P.A. and Morgan, R.A. (1987) Diet and feeding ecology of Procellariiformes. In Croxall, J.P. (ed.), *Seabirds: Feeding Ecology and Role in Marine Ecosystems*, Cambridge University Press, Cambridge, 135–171.
Rudels, B. (1987) On the mass balance of the Polar Ocean, with special emphasis on the Fram Strait. *Norsk Polarinstitutt Skrifter* **188**.
Sage, B. (ed.) (1986) *The Arctic and its Wildlife*. Croom Helm, London.
Siegfried, W.R. (1985) Birds and mammals—oceanic birds of the Antarctic. In Bonner, W.N. and Walton, D.W.H. (eds.), *Antarctica*, Pergamon, Oxford, 242–265.
Stirling, I. and Cleator, H. (ed.) (1981) Polynyas in the Canadian Arctic. *Canadian Wildlife Service Papers* **45**.
Stonehouse, B. (1967) Occurrence and effects of open water in McMurdo Sound during winter and early spring. *Polar Record* **13**(87), 775–778.
Stonehouse, B. (1971) *Animals of the Arctic*. Ward Locke, London.
Stonehouse, B. (1972) *Animals of the Antarctic*. Peter Lowe, London.
Stonehouse, B. (1985a) *Sea Mammals of the World*. Penguin, London.
Stonehouse, B. (1985b) Birds and mammals—Penguins. In Bonner, W.N. and Walton, D.W.H. (eds.), *Antarctica*, Pergamon, Oxford, 266–292.
Swithinbank, C.W.M. and others (1977) Drift tracks of Antarctic icebergs. *Polar Record* **18**(116), 495–501.
Tchernia, P. and Jeannin, P.F. (1984) Circulation in antarctic waters as revealed by iceberg tracks 1972–83. *Polar Record* **22**(138), 263–269.
Tonessen, J.N. and Johnsen, A.O. (1982) *The History of Modern Whaling*. Hurst, London.
Voronina, N.M. (1966) The distribution of zooplankton biomass in the Southern Ocean. *Oceanology* **6**, 836–846.
Voronina, N.M. (1968) The distribution of zooplankton in the Southern Ocean and its dependence on the circulation of water. *SARSIA* **34**, 277–284.
Watson, G.E. (1975) *Birds of the Antarctic and Sub-Antarctic*. American Geophysical Union, Washington DC.
Young, O.R. (1981) The political economy of the northern fur seal. *Polar Record* **20**(128), 407–416.
Zenkevitch, L.A. (1963) *Biology of the Seas of the USSR*. Allen and Unwin, London.

Chapter 6

Addison, P.A. (1977) Studies on evapotranspiration and energy budgets on Truelove Lowland. In Bliss, L.C. (ed.), *Truelove Lowland, Devon Island, Canada: a High Arctic Ecosystem*, University of Alberta Press, Edmonton, 281–300.
Bliss, L.C. (1971) Arctic and alpine plant life cycles. *Ann. Rev. Ecol. and Systematics* **2**, 405–438.
Block, W. (1977) Oxygen consumption of the terrestrial mite *Alaskozetes antarcticus* (Acari: Cryptostigmata). *J. Experimental Biol.* **68**, 69–87.
Bonner, W.N. (1985) Impact of fur seals on the terrestrial environment at South Georgia. In

Siegfried, W.R., Condy, P.R. and Laws, R.M. (eds.), *Antarctic Nutrient Cycles and Food Webs*, Springer, Berlin, 641–646.

Chernov, Yu. I. (1975) *Prirodnaya zonal'nost' i zhivotnyy mir sushi* [Natural zonation and animal life on the continent]. Moscow. (In Russian.)

Chernov, Yu.I. (1978) Adaptive features of the life cycles of tundra zone insects. *Zhurnal Obshchoi Biologia* **39**, 394–401. (In Russian.)

Chernov, Yu.I. (1985) *The Living Tundra.* Cambridge University Press, Cambridge.

Courtin, G.M. and Mayo, J.M. (1975) Arctic and alpine plant water relations. In Vernberg, F.J. (ed.), *Physiological Adaptation to the Environment.* Intext Educational, New York, 201–228.

Collins, N.J. (1977) The growth of mosses from two contrasting communities in the maritime Antarctic: measurement and prediction of net annual production. In Llano, G.A. (ed.), *Adaptations within Antarctic Ecosystems*, Smithsonian Institution, Washington DC, 921–933.

Danks, H.V. (1978) Modes of seasonal adaptation in the insects. I. Winter survival. *Can. Entomologist* **110**, 1116–1205.

Danks, H.V. (1981) *Arctic Arthropods: A Review of Systematics and Ecology with Particular Reference to the North American Fauna.* Entomological Society of Canada, Ottawa.

DeVries, A.L. (1978) The physiology and biochemistry of low temperature adaptations in polar marine ectotherms. In McWhinnie, M.A. (ed.), *Polar Research. To the Present, and the Future. AAAS Selected Symp.* **7**, 175–202.

Dodge, C.W. (1973) *Lichen Flora of the Antarctic Continent and Adjacent Islands.* Phoenix, Canaan, NH.

Edwards, J.A. (1972) Studies in *Colobanthus quitensis* (Kunth) Bartl. and *Deschampsia antarctica* Desv.: V. Distribution, ecology and vegetative performance on Signy Island. *Bull. Br. Antarctic Surv.* **28**, 11–28.

Edwards, J.A. (1973) Vascular plant production in the maritime Antarctic. In Bliss, L.C. and Wielgolaski, F.E. (eds.), *Primary Production and Production Processes, Tundra Biome.* Stockholm, Tundra Biome Steering Committee, Stockholm, 169–175.

Edwards. J.A. (1974) Studies in *Colobanthus quitensis* (Kunth) Bartl. and *Deschampsia antarctica* Desv.: VI. Reproductive performance on Signy Island. *Bull. Br. Antarctic Surv.* **39**, 67–86.

Edwards, J.A. (1975) Studies in *Colobanthus quitensis* (Kunth) Bartl. and *Deschampsia antarctica* Desv.: VII. Cyclic changes related to age in *Colobanthus quitensis. Bull. Br. Antarctic Surv.* **40**, 1–6.

Edwards, J.A. (1980) An experimental introduction of vascular plants from South Georgia to the maritime Antarctic. *Bull. Br. Antarctic Surv.* **49**, 73–80.

Everson, I. (1984) Fish biology. In Laws, R.M. (ed.), *Antarctic Ecology* 2. Academic Press, London, 491–532.

Everson, I. (1987) Physiological adaptations. In Walton, D.W.H. (ed.), *Antarctic Science*, Cambridge University Press, Cambridge, 97–112.

Gannutz, T.P. (1969) Effects of environmental extremes on lichens. *Bull. Soc. Botanique de France, Mémoires* 1968.

Gannutz, T.P. (1970) Photosynthesis and respiration in of plants in the Antarctic Peninsula area. *Antarctic J. of the United States* **5**, 49–51.

Gusta, L.V. (1985) Freezing resistance in plants. In Kaurin, Å, Junttila, O. and Nilsen, J. (eds.), *Plant Production in the North*, Norwegian University Press, Oslo, 219–235.

Haber, E. (1986) Flora of the circumpolar Arctic. In Sage, B. (ed.), *The Arctic and its Wildlife*, Croom Helm, Beckenham, 59–71.

Hargens, A.R. (1972) Freezing resistance in polar fishes. *Science* **176**, 184–186.

Hedberg, O. (1964). *Features of Afroalpine Plant Ecology.* Almqvist and Wikselles, Uppsala.

Heide, O.M. (1984) Physiological aspects of climatic adaptation in plants with special reference to high-latitude environments. In Kaurin, Å, Junttila, O. and Nilsen, J. (eds.), *Plant Production in the North*, Norwegian University Press, Oslo, 1–22.

Heinen, S. and Karunen, P. (1985) Seasonal variation of net photosynthesis and galactolipids in *Dicranum elongatum*. In Kaurin, Å, Junttila, O. and Nilsen, J. (eds.), *Plant Production in the North*, Norwegian University Press, Oslo, 246–253.
Heywood, R.B. (1977) Antarctic freshwater ecosystems: review and synthesis. In Llano, G.A. (ed.), *Adaptations within Antarctic Ecosystems*, Smithsonian Institution, Washington DC, 801–828.
Heywood, R.B. (1984) Antarctic inland waters. In Laws, R.M. (ed.), *Antarctic Ecology* **1**, Academic Press, London, 279–344.
Holmen, K. (1960) The mosses of Peary Land, North Greenland. *Meddellelser om Grønland* **124**(9), 1–149.
Kacperska, A. (1984) Biochemical and physiological aspects of frost hardening in herbaceous plants. In Kaurin, Å, Junttila, O. and Nilsen, J. (eds.), *Plant Production in the North*, Norwegian University Press, Oslo, 99–115.
Kanwisher, J. (1959) Histology and metabolism of frozen intertidal animals. *Biol. Bull.* **116**, 258–264.
Kappen, I. and Friedmann, E.I. (1983) Ecophysiology of lichens in the Dry Valleys of southern Victoria Land, Antarctica. II. Co_2 gas exchange in cryptoendolithic lichens. *Polar Biology* **1**(4), 227–232.
Kappen, I. and Lange, O.L. (1972) Die Kälteresistenze einiger Makrolichen. *Flora* **161**, 129.
Kappen, I., Friedmann, E.I. and Garty, J. (1981) Ecophysiology of lichens in the Dry Valleys of southern Victoria Land, Antarctica: 1. Microclimate of the cryptoendolithic lichen habitat. *Flora* **171**, 216–235.
Kaurin, Å (1984) Effects of light quality on frost hardening in *Poa alpina*. In Kaurin, Å, Junttila, O. and Nilsen, J. (eds.), *Plant Production in the North*, Norwegian University Press, Oslo, 116–126.
Kaurin, Å, Junttila, O. and Nilsen, J. (eds.) (1984) *Plant Production in the North*. Norwegian University Press, Oslo.
Kevan, P.G. and Danks, H.V. (1986*a*) Adaptations of arctic insects. In Sage, B. (ed.), *The Arctic and its Wildlife*, Croom Helm, Beckenham, 55–57.
Kevan, P.G. and Danks, H.V. (1986*b*) Arctic insects. In Sage, B. (ed.), *The Arctic and its Wildlife*, Croom Helm, Beckenham, 72–77.
Kleiber, M. (1961) *The Fire of Life: An Introduction to Animal Energetics*. Wiley, New York.
Krog, J. (1955) Notes on temperature measurements indicative of special organization in arctic and subarctic plants for utilization of radiated heat from the sun. *Physiol. Pl.* **8**, 836–839.
Lamb, I.M. (1970) Antarctic terrestrial plants and their ecology. In Holdgate, M.W. (ed.), *Antarctic Ecology* **2**, Academic Press, London, 733–751.
Lange, O.L. and Kappen, L. (1972) Photosynthesis of lichens from Antarctica. In Llano, G.A. (ed.), *Antarctic Terrestrial Biology. Antarctic Research Series* **20**, American Geophysical Union, Washington DC, 83–95.
Larsen, J.A. (1964) The role of physiology and environment in the distribution of Arctic plants. *University of Wisconsin Technical Rept.* **16**, University of Wisconsin, Madison.
Lin, Y., Duman, J.G. and DeVries, A.L. (1972) Studies on the structure and activity of low molecular weight glycoproteins from an antarctic fish. *Biochem. Biophys. Res. Comm.* **46**, 87–92.
Lindsey, C.C. (1981) Arctic refugia and the evolution of arctic biota. In Scudder, G.E. and Reveal, J.L. (eds.), *Evolution Today, Proc. Second Int. Congr. of Systematic and Evolutionary Biology*, Carnegie-Mellon University, Pittsburgh, 7–10.
Longton, R.E. (1985) Terrestrial habitats—vegetation. In Bonner, W.N. and Walton, D.W.H. (eds.), *Antarctica*, Pergamon, Oxford, 73–105.
Löve, A., and Löve, D. (1975) *Cytotaxonomical Atlas of the Arctic Flora*. J. Cramer, Vaduz.
Mayo, J.M., Hartgerink, A.P., Despain, D.D., Thompson, R.G., van Zinderen Bakker, E.M. and Nelson, S.D. (1977) Gas exchange studies of *Carex, Dryas*, Trulove Lowland. In Bliss, L.C. (ed.), *Truelove Lowland, Devon Island, Canada: a High Arctic Ecosystem*, University of Alberta Press, Edmonton, 265–280.

Mooney, H.A. and Billings, W.D. (1961) Comparative physiological ecology of arctic and alpine populations of *Oxyria digyna*. *Ecol. Monogr.* **31**, 1–29.
Mooney, H.A. and Johnson, A.W. (1965) Comparative physiological ecology of an arctic and an alpine population of *Thalictrum alpinum* L. *Ecology* **46**, 721–727.
Mosquin, T. (1966) Reproductive specialization as a factor in the evolution of the Canadian flora. In Taylor, R.L. and Ludwig, R.A. (eds.), *The Evolution of Canada's Flora*, University of Toronto Press, Toronto, 41–63.
Procter, D.L.C. (1977) Nematode densities and production on Truelove Lowland. In Bliss, L.C. (ed.), *Truelove Lowland, Devon Island, Canada: a High Arctic Ecosystem*, University of Alberta Press, Edmonton, 347–361.
Ryan, J.K. (1977) Synthesis of energy flows and population dynamics of Truelove Lowland invertebrates. In Bliss, L.C. (ed.), *Truelove Lowland, Devon Island, Canada: a High Arctic Ecosystem*, University of Alberta Press, Edmonton, 325–346.
Sage, B. (1986) *The Arctic and its Wildlife*. Croom Helm, Beckenham, 78–117.
Salisbury, F.B. (1985) Plant adaptations to the light environment. In Kaurin, Å, Junttila, O. and Nilsen, J. (eds.), *Plant Production in the North*. Norwegian University Press, Oslo, 43–61.
Savile, D.B.O. (1960) Limitations of the competitive exclusion principle. *Science* **132**, 1761.
Savile, D.B.O. (1968) Land plants. In Beals, C.S. (ed.), *Science, History and Hudson Bay* **1**, Department of Energy, Mines and Resources, Ottawa, 397–416.
Savile, D.B.O. (1972) *Arctic Adaptations in Plants. Monograph* **6**, Canada Department of Agriculture, Ottawa.
Savile, D.B.O. and Calder, J.A. (1952) Notes on the flora of Chesterfield Inlet, Keewatin District, NWT. *Canadian Field Naturalist* **66**, 103–107.
Schmidt-Nielsen, K. (1975) *Animal Physiology*. Cambridge University Press, Cambridge.
Schmidt-Nielsen, K. (1976) *How Animals Work*. Cambridge University Press, Cambridge.
Scholander, P.F., Flagg, W., Hock, R.J. and Irving, L. (1953) Studies on the physiology of frozen plants and animals in the arctic. *J. Cell. Compar. Physiol.* **42**, supplement 1, 1–56.
Scholander, P.F., Van Dam, L., Kanwisher, J.W., Hammel, H.T. and Gordon, M.S. (1957) Supercooling and osmoregulation in arctic fish. *J. Cell. Compar. Physiol.* **49**, 5–24.
Siegfried, W.R. (1985) Birds and mammals—oceanic birds of the Antarctic. In Bonner, W.N. and Walton, D.W.H. (eds.), *Antarctica*, Pergamon, Oxford, 242–265.
Smith, R.I.L. (1984) Terrestrial plant ecology of the sub-Antarctic and Antarctic. In R.M. Laws (ed.), *Antarctic Ecology* **1**, Academic Press, London, 61–162.
Smith, R.I.L. and Poncet, S. (1985) New southernmost record for Antarctic flowering plants. *Polar Record* **22** (139) 425–427.
Stonehouse, B. (1967) The general biology and thermal balances of penguins. In Cragg, J.B. (ed.) *Advances in Ecological Research* **4**, Academic Press, London, 131–196.
Stonehouse, B. (1985) Birds and mammals—penguins. In Bonner, W.N. and Walton, D.W.H. (eds.), *Antarctica*, Pergamon, Oxford, 266–292.
Tronsmo, A.M. and Kaurin, Å. (1985) Adaptation to low-temperature stress. In Kaurin, Å, Junttila, O. and Nilsen, J. (eds.), *Plant Production in the North*, Norwegian University Press Oslo, 332–333.
Warren Wilson, J. (1964) Annual growth of *Salix arctica* in the high arctic. *Ann. Bot.* **28**, 71–76.
Whitfield, D.W.A. and Goodwin, C.R. (1977) Comparison of the estimates of annual vascular plant production on Truelove Lowland made by harvesting and by gas exchange. In Bliss, L.C. (ed.), *Truelove Lowland, Devon Island, Canada: a High Arctic Ecosystem*, University of Alberta Press, Edmonton, 315–321.

Chapter 7

Armstrong, T.E., Rogers, G. and Rowley, G. (1978) *The Circumpolar North*. Methuen, London.

Cook, J. (1961) The voyage of the Resolution and Adventure (1772–75) (ed. J.C. Beaglehole), Hakluyt Society, London.
Birket-Smith, K. (1959) *The Eskimos.* Methuen, London.
Bonner, W.N. (1980) *Whales.* Blandford Press, Poole.
Bonner, W.N. (1984) Introduced mammals. In Laws, R.M. (ed.), *Antarctic Ecology* **1**, Academic Press, Oxford, 237–278.
Brotherhood, J. (1973) Studies on energy expenditure in the Antarctic. In Edholm, O.G. and Gunderson, E.K.E. (eds.), *Polar Human Biology*, Heinemann, London, 182–192.
Dalrymple, P.C. (1966) A physical climatology of the Antarctic plateau. In M.J. Rubin (ed.), *Studies in Antarctic Meteorology, Antarctic Res. Ser.* **9**, 195–231.
Durnin, J.V.G.A. and Passmore, R. (1967) *Energy, Work and Leisure.* Heinemann, London.
Eddie, G.O. (1977) The harvesting of krill. *FAO Rept* GLO/SO/77/2, Food and Agriculture Organization of the United Nations, Rome.
Edholm, O.G. (1978) *Man—Hot and Cold.* Edward Arnold, London.
Everson, I. (1977) The living resources of the Southern ocean. *FAO Rept* GLO/SO/77/1, Food and Agriculture Organization of the United Nations, Rome.
Gjelsvik, T. (1983) The mineral resources of Antarctica: progress in their identification. In Orrego Vicuna, F. (ed.), *Antarctic Resources Policy*, Cambridge University Press, Cambridge, 61–76.
Godin, G. and Shephard, R.J. (1973) Activity patterns of the Canadian Eskimo. In Edholm, O.G. and Gunderson, E.K.E. (eds.), *Polar Human Biology,* Heinemann, London, 193–215.
Grantham, G.J. (1977) The utilization of krill. *FAO Rept* GLO/SO/77/3, Food and Agriculture Organization of the United Nations, Rome.
Hall, S. (1987) *The Fourth World; the Heritage of the Arctic and its Destruction.* Bodley Head, London.
Hayes, G. (1932) *Conquest of the South Pole. Antarctic Exploration 1906–1931.* Butterworth, London.
Herbert, W. (1976) *Eskimos.* Collins, London.
Holdgate, M.W. and Tinker, J. (1979) Oil and other minerals in the Antarctic. Scientific Committee on Antarctic Research, Cambridge.
Hopkins, D.M. (ed.) (1967) *The Bering Land Bridge.* Stanford University Press, Stanford.
Ignatov, V.S. (1965) At the pole of cold. *Priroda* **9**, 74–76 (In Russian).
Knox, G.A. (1983) The living resources of the southern ocean: a scientific overview. In Orrego Vicuna, F. (ed.), *Antarctic Resources Policy,* Cambridge University Press, Cambridge, 21–60.
Mill, H.R. (1905) *The Siege of the South Pole.* Alston Rivers, London.
Orrego Vicuna, F. (1983) *Antarctic Resources Policy.* Cambridge University Press, Cambridge.
Parsons, A. (1987) *Antarctica: the Next Decade.* Cambridge University Press, Cambridge.
Reader's Digest (1985) *Antarctica: Great stories from the Frozen Continent.* Reader's Digest Association, London.
Siple, P.A. and Passel, C.F. (1945) Measurements of dry atmospheric cooling in sub-freezing temperatures. *Proc. Amer. Phil. Soc.* **89**, 177–199.
Sheldon, J.F. (1988) Oil versus caribou in the Arctic; the great debate. *Polar Record* **24** (149) 95–100.
Shephard, R.J. and Rode, A. (1973) Fitness for Arctic life: the cardio-respiratory status of the Canadian Eskimo. In Edholm, O.G. and Gunderson, E.K.E. (eds.), *Polar Human Biology*, Heinemann, London, 216–245.
Stonehouse, B. (1971) *Animals of the Arctic.* Ward Lock, London.
Triggs, G.D. (1987) *The Antarctic Treaty Regime: Law, Environment and resources.* Cambridge University Press, Cambridge.
Theutenberg, B.J. (1984) *The Evolution of the Law of the Sea.* Tycooly International, Dun Laoghaire.

Index